HAESE MATHEMATICS

Specialists in mathematics education

Mathematics 3

PYP 3 Write-on Workbook

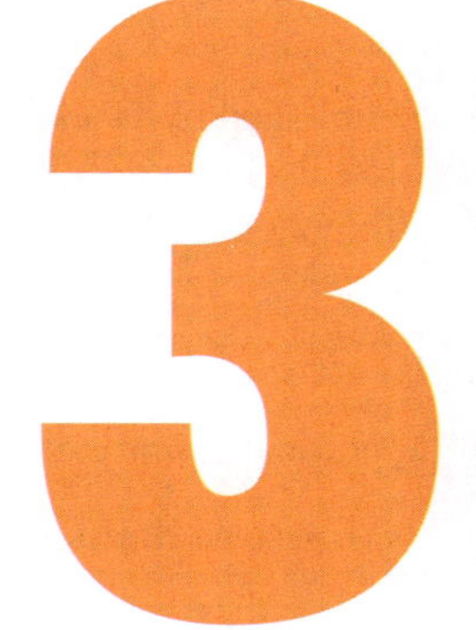

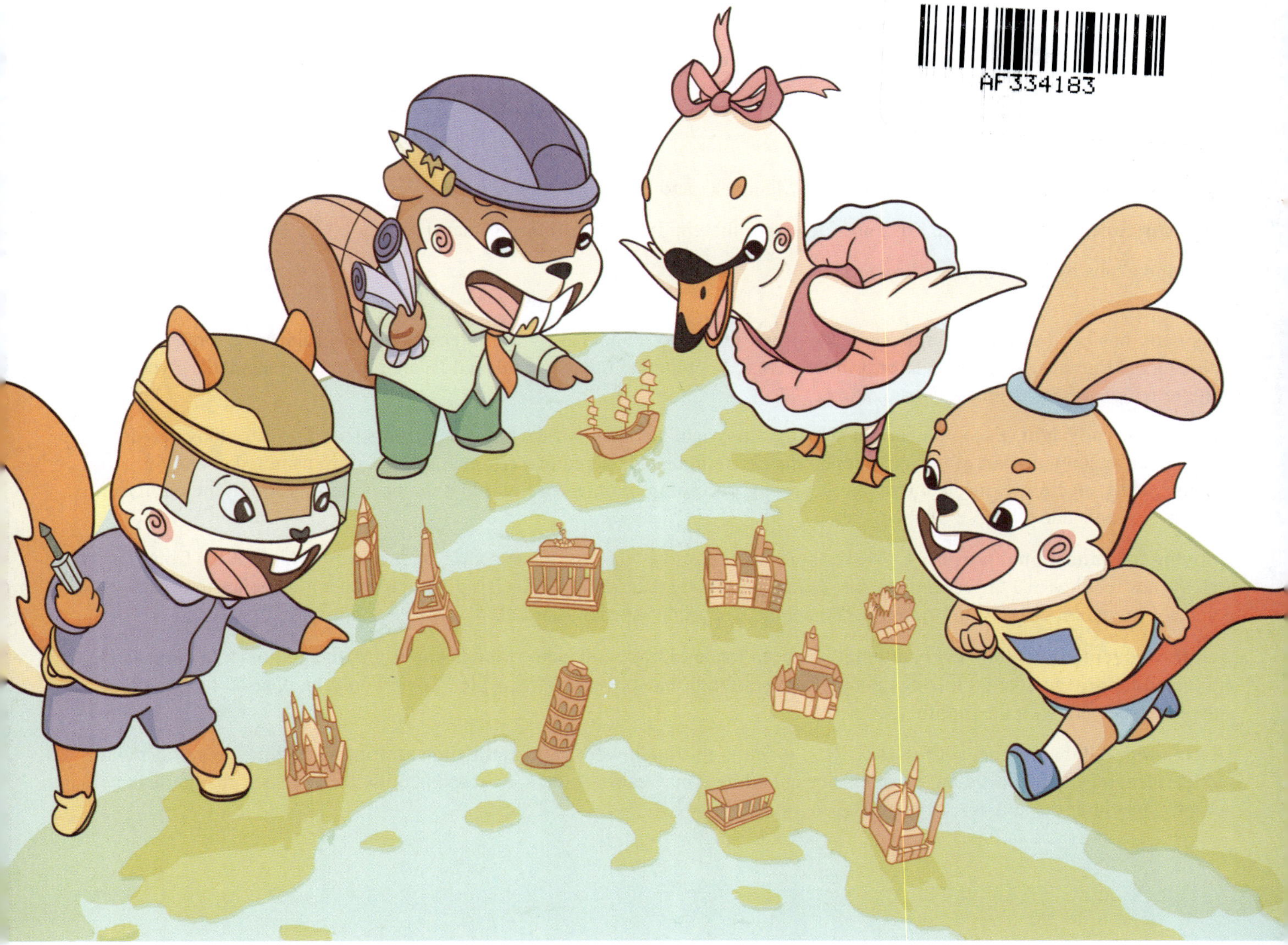

for use with

IB Primary Years Programme

Michael Haese   Mark Humphries   Philippa Sawyers

MATHEMATICS PYP 3

Michael Haese   B.Sc.(Hons.), Ph.D.
Mark Humphries   B.Sc.(Hons.)
Philippa Sawyers   B.Ed.(Hons.)

Published by Haese Mathematics
152 Richmond Road, Marleston, SA 5033, AUSTRALIA
Telephone: +61 8 8210 4666
Email: info@haesemathematics.com
Web: www.haesemathematics.com

National Library of Australia Card Number & ISBN 978-1-922416-77-3

© Haese & Harris Publications 2024

First Edition   2024

Cartoon artwork by James Hobbs, Yi-Tung Huang, and Le Quynh Chi Ta.

Artwork by Hannah Coleman and Brian Houston.

Cover art by Le Quynh Chi Ta.

Computer software by Patrick French, Ben Hensley, Huda Kharrufa, Rachel Lee, and Han Zong Ng.

Audio recorded by Eloise Quinn-Valentine.

Production work by Hannah Coleman, Michael Mampusti, and Ngoc Vo.

Typeset in Australia by Charlotte Frost and Deanne Gallasch. Typeset in Times Roman 13.

Printed in China by Prolong Press Limited.

FOREWORD

Mathematics PYP 3 has been designed and written for the International Baccalaureate Primary Years Programme (IB PYP). The workbook and online interactive material provide an engaging and structured package, allowing students to explore and develop their confidence in Mathematics.

Our aim with this spiral bound write-on workbook is to provide the students and teachers with an alternative to photocopied worksheets, which in turn will give the students an organised record of their work throughout the year.

The book contains a variety of exercises ranging from basic to advanced, to cater for a range of student abilities and interests. The material is presented in a clear, easy-to-follow style to aid comprehension and retention, especially for English Language Learners. Each chapter ends with a set of Revision questions.

Important information and key notes are highlighted, and worked examples provide clear instruction and relevant explanations. Discussions, Activities, and Puzzles are used throughout the chapters to develop understanding. All of these features can be viewed through Snowflake, our online digital textbook platform.

We have endeavoured to provide a stimulating workbook and accompanying interactive material. Our aim is to develop and encourage student understanding and to grow an appreciation and love for Mathematics.

We welcome your feedback:

Email: info@haesemathematics.com
Web: www.haesemathematics.com

ONLINE FEATURES

Each workbook comes with a 12 month subscription to the online edition and its range of interactive features. This can be accessed through the **SNOWFLAKE** online learning platform via a web browser or our offline viewer.

To activate your electronic workbook, please contact Haese Mathematics by emailling info@haesemathematics.com with your proof of purchase such as a copy of your receipt.

For general queries regarding **SNOWFLAKE** and online subscriptions:

- Visit our help page: https://snowflake.haesemathematics.com.au/help
- Contact Haese Mathematics: info@haesemathematics.com

WATCH LISTEN LEARN

Watch Listen Learn is an exciting feature of this book.

This icon in a worked example or exercise denotes an active online link.

Simply click the icon (or anywhere in the example or exercise box) to access the Watch Listen Learn with a teacher's voice explaining each concept.

Play any line as often as you like. See how the basic processes come alive using movement and colour on the screen.

CLICKABLE ICONS

You will see these icons throughout the workbook. Click the icon to access online content.

Printable Worksheets containing extra questions or support material.

Listening Activities.

Video Demonstrations providing instruction.

Activities and Games.

TABLE OF CONTENTS

CHAPTER 1: NUMBER

The symbols 0, 1, 2, 3, 4, 5, 6, 7, 8, and 9 are called **digits**.

We name each digit by the number it represents.

0	zero	5	five
1	one	6	six
2	two	7	seven
3	three	8	eight
4	four	9	nine

Zero is used when there are no objects to count.

We can draw other small counting numbers using **unit** blocks.

Exercise 1

Practise your spelling.

digit ________________________   unit ________________________

Copy each number.

0	zero ________	5	five ________
1	one ________	6	six ________
2	two ________	7	seven ________
3	three ________	8	eight ________
4	four ________	9	nine ________

A **numeral** is a number written using digits.

Once we count past nine, we use more than one digit to write each number in numeral form.

For example, the number after nine is ten, written 10.

Exercise 2

Diagram	Numeral	Word
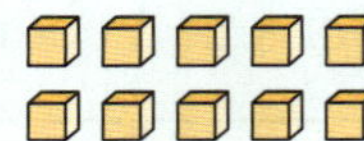	10	ten

Practise your spelling:

numeral _______________________________   ten _______________________________

How many digits are used to write the numeral 10? _______________

This block is a **ten**.   It represents 10 units.

Exercise 3

Practise your spelling.

11   eleven

____

14   fourteen

____

17   seventeen

____

12   twelve

____

15   fifteen

____

18   eighteen

____

13   thirteen

____

16   sixteen

____

19   nineteen

____

Discussion

Why do you think we write numbers in numeral form using only the ten different digits?

Exercise 4

Practise your spelling.

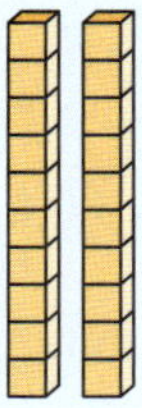

20 twenty

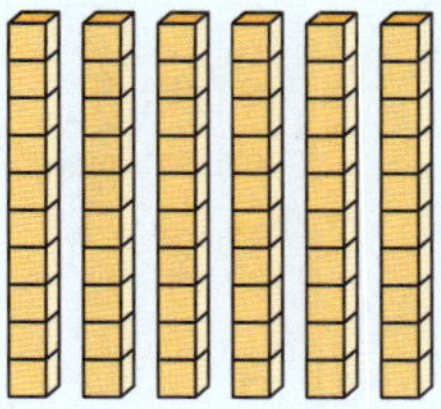

____

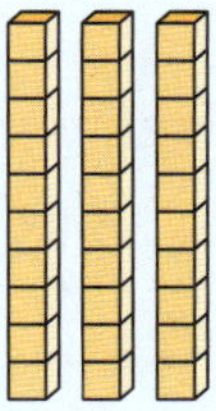

30 thirty

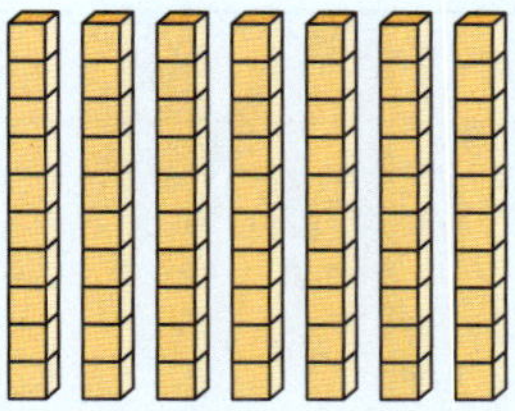

____

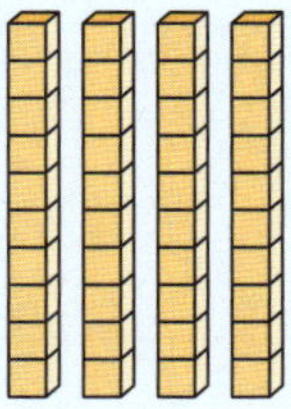

40 forty

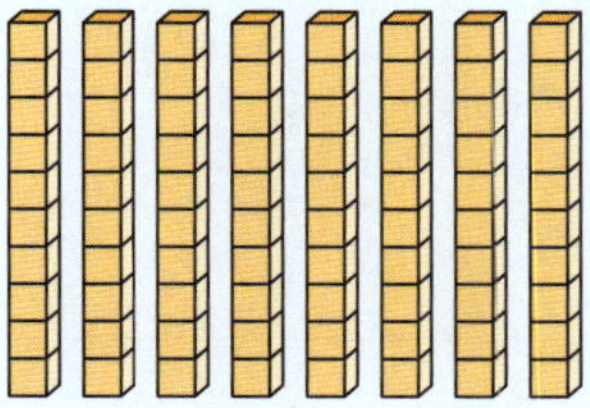

____

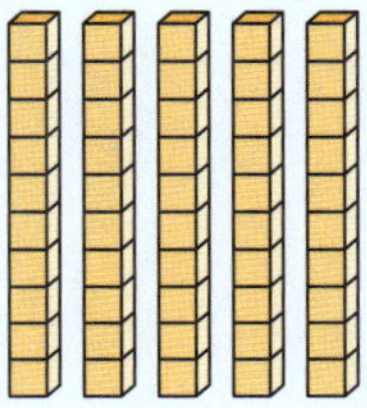

50 fifty

____

60 sixty

____

70 seventy

____

80 eighty

____

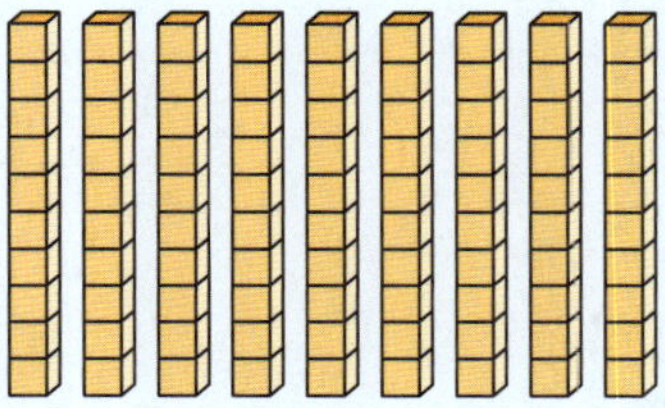

90 ninety

____

Listening Activity

Click to practise listening to and saying numbers.

Listening Activity

Click and follow the instructions.

1 ________________   2 ________________   3 ________________

4 ________________   5 ________________   6 ________________

7 ________________   8 ________________   9 ________________

10 ________________   11 ________________   12 ________________

We use the symbols $=$ and $\neq$ to *compare* numbers.

"equal to"   "not equal to"

Exercise 5

Practise your spelling.

equals ________________   equal to ________________

________________   ________________

Exercise 6

Complete using $=$ or $\neq$.

a  ______

b 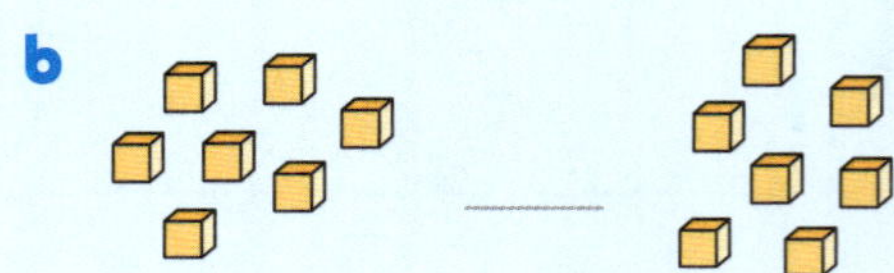 ______

c 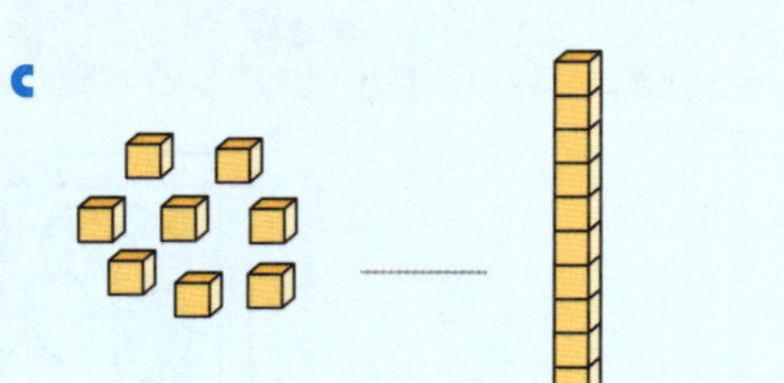 ______

d 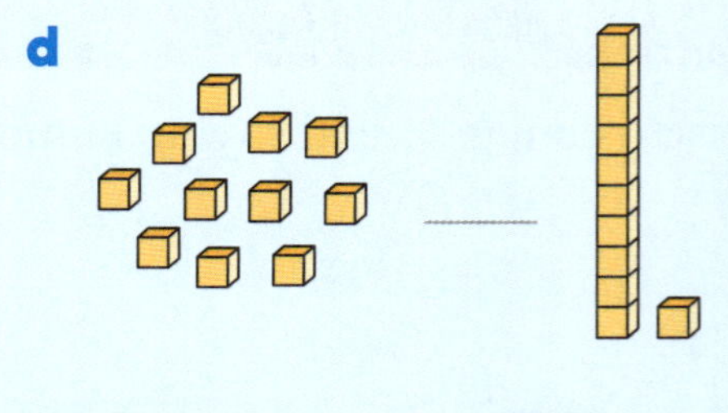 ______

e 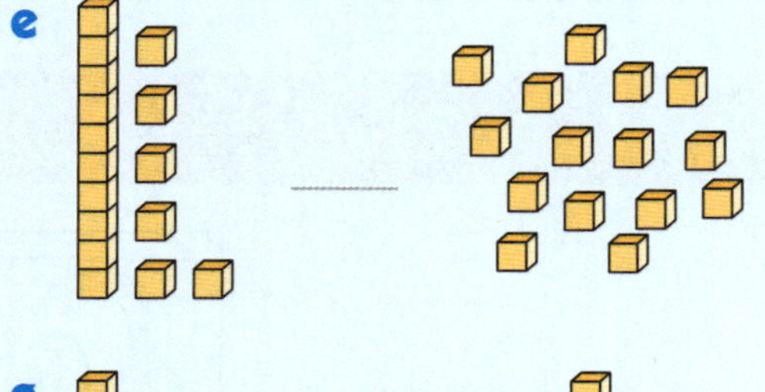 ______

f 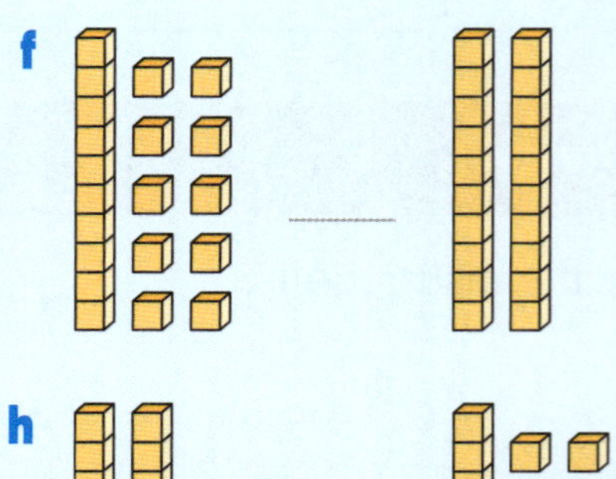 ______

g 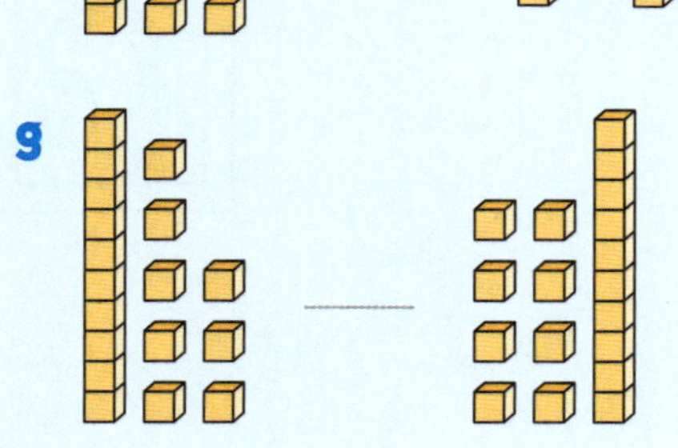 ______

h  ______

> We write our numbers using **place values**.
> **Tens** and **units** are place values.

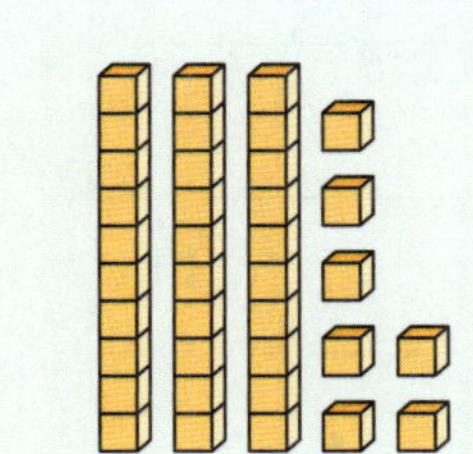

3 tens and 7 units

$$30 \quad + \quad 7 \quad = \quad 37$$

thirty seven

Exercise 7

What numbers are shown?

a   ☐ tens and ☐ units   ☐ + ☐ = ☐

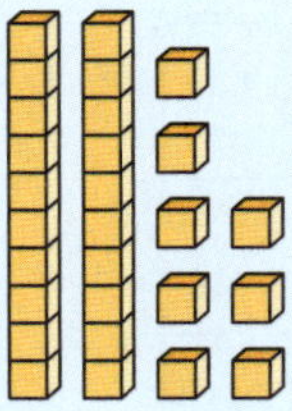

b   ☐ tens and ☐ units   ☐ + ☐ = ☐

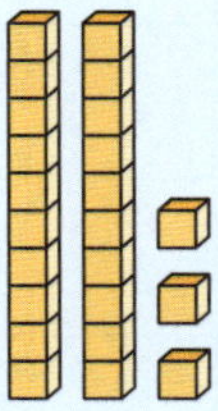

c   ☐ tens and ☐ units   ☐ + ☐ = ☐

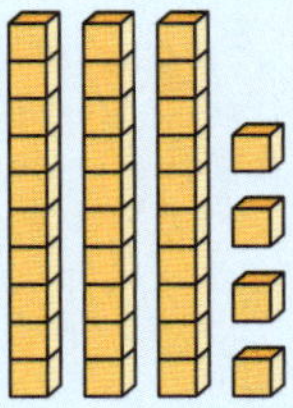

d   ☐ tens and ☐ units   ☐ + ☐ = ☐

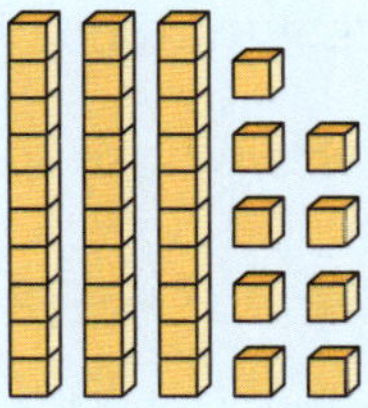

e   ☐ tens and ☐ units   ☐ + ☐ = ☐

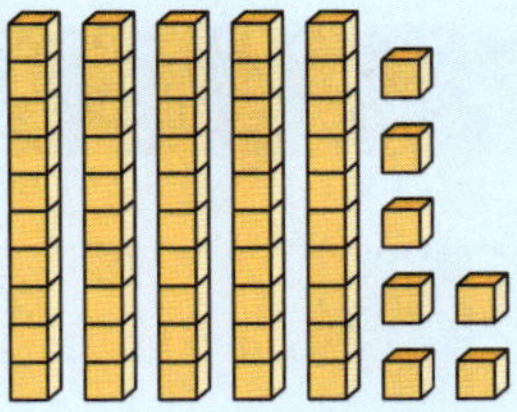

f   ☐ tens and ☐ units   ☐ + ☐ = ☐

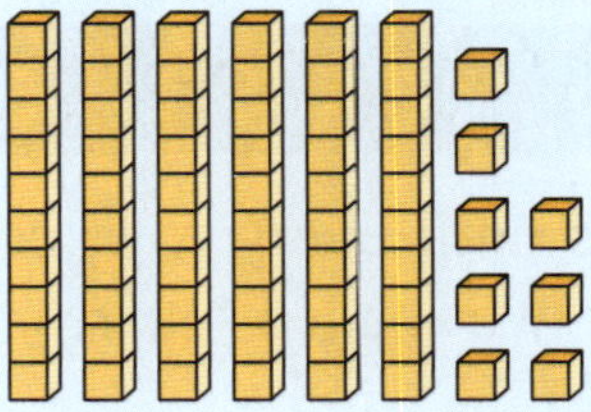

Activity

1 Complete:

	2	3		5				9	
11			14			17			
	22				26				30
		34				37			
	43		45				48		
51					56				60
	62		64					69	
71						77			
	82			85				89	
		93			96		98		100

2 Colour in *yellow* the numbers written with only one digit.

3 Colour in *red* the numbers with tens but no units.

4 Colour in *green* the numbers whose digits are the same.

5 Colour in *blue* the numbers whose digits are 2 and 6.

6 Colour in *orange* the numbers whose digits are 4 and 8.

7 Colour in *purple* the numbers whose digits are 3 and 5.

Puzzle

Shade more squares until every row, column, and blue diagonal has *exactly* two shaded squares.

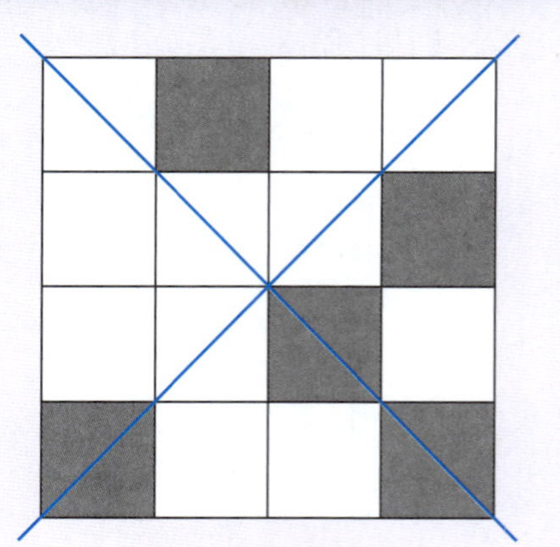

Exercise 8

Partition each number using place values:

3   9

3 tens   9 units

$30 + 9$

thirty nine

a   6   8

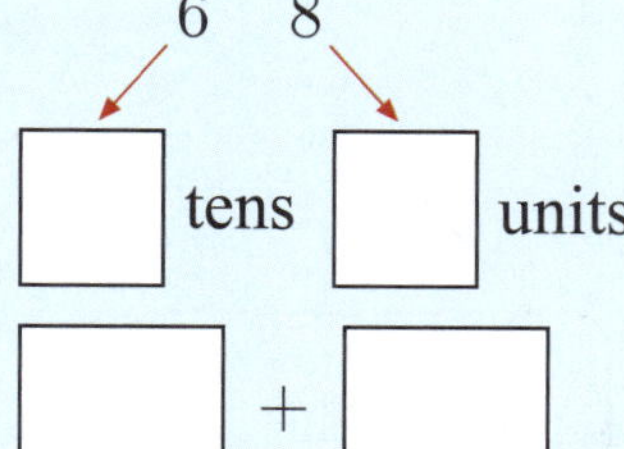

☐ tens   ☐ units

☐ + ☐

b   5   1

☐ tens   ☐ unit

☐ + ☐

c   4   7

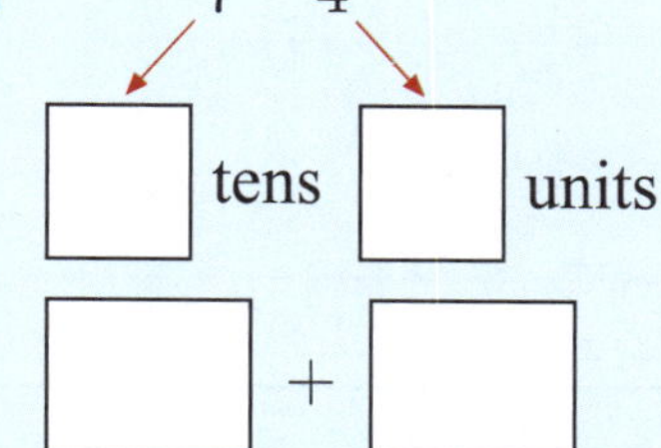

☐ tens   ☐ units

☐ + ☐

d   7   4

☐ tens   ☐ units

☐ + ☐

e   9   6

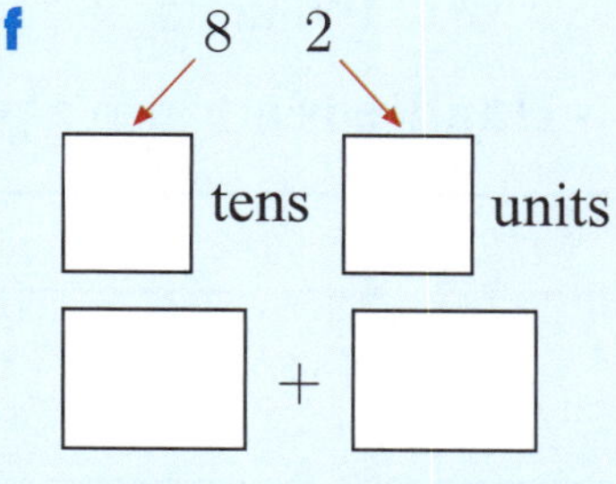

☐ tens   ☐ units

☐ + ☐

f   8   2

☐ tens   ☐ units

☐ + ☐

Listening Activity

Click and follow the instructions.

1 _______________

2 _______________

3 _______________

4 _______________

5 _______________

6 _______________

Exercise 9

Write in words:

a 28 _______________________________

b 34 _______________________________

c 49 _______________________________

d 51 _______________________________

e 65 _______________________________

f 72 _______________________________

g 86 _______________________________

h 93 _______________________________

i 47 _______________________________

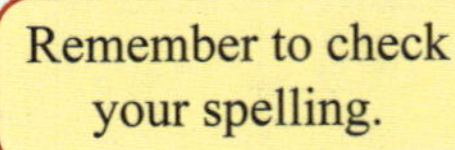

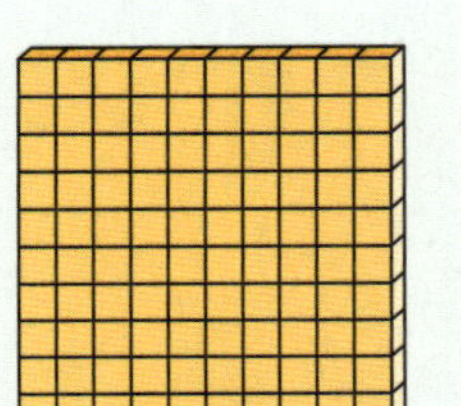
This block is a **hundred**.

It represents 10 tens

   or   100 units

Hundreds are also a **place value**.
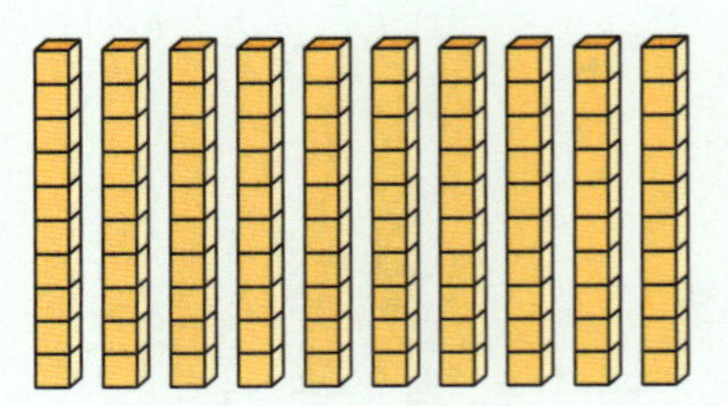

Exercise 10

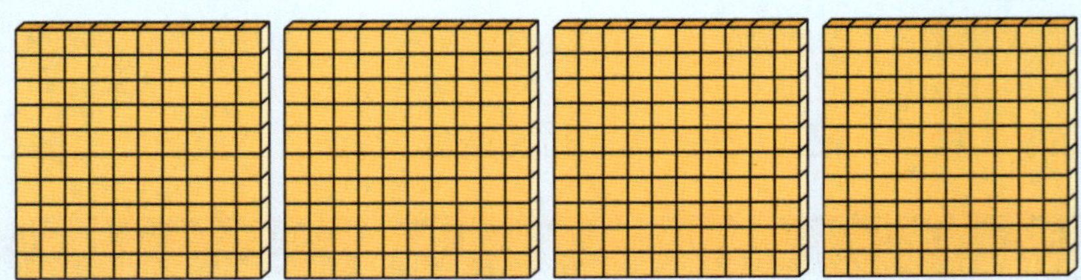

a How many hundreds are shown? _______________

b Write the number as a numeral. _______________

c Write the number in words. _______________________________________

Exercise 11

What number is shown?

1 hundred   2 tens   4 units

The number is 124 or one hundred and twenty four.

a

☐ hundred   ☐ tens   ☐ units

The number is ☐

or ___________________________________.

b

☐ hundreds

☐ tens

☐ units

The number is ☐

or ___________________________________.

c

☐ hundreds   ☐ tens   ☐ units

The number is ☐

or ___________________________________.

Exercise 12

Partition each number using place values:

a

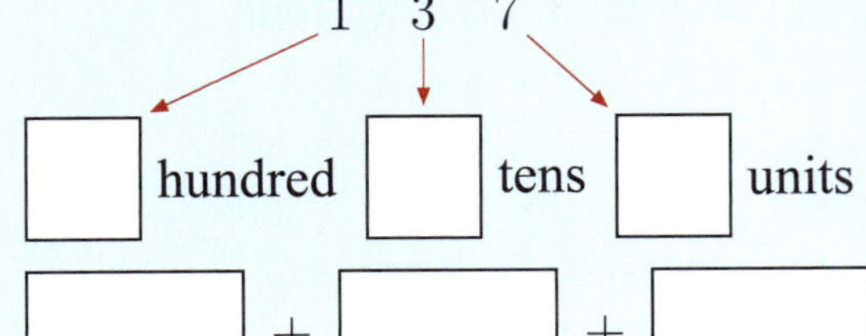

b

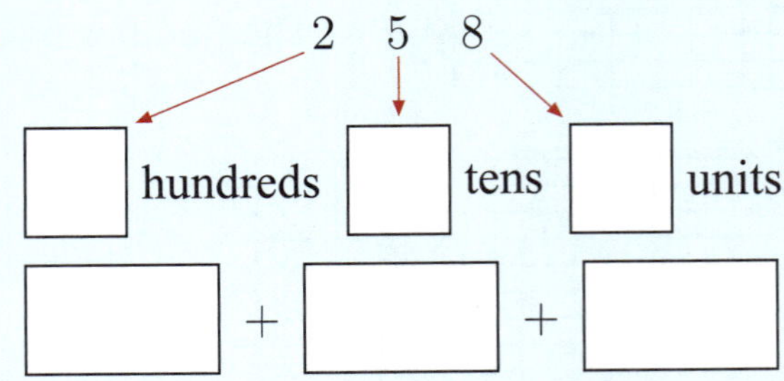

c

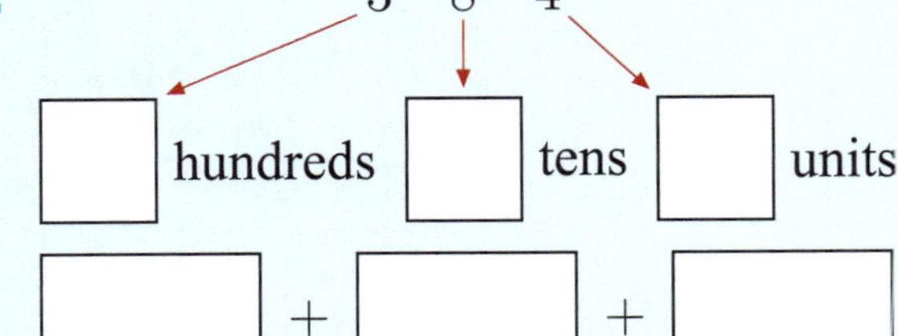

d

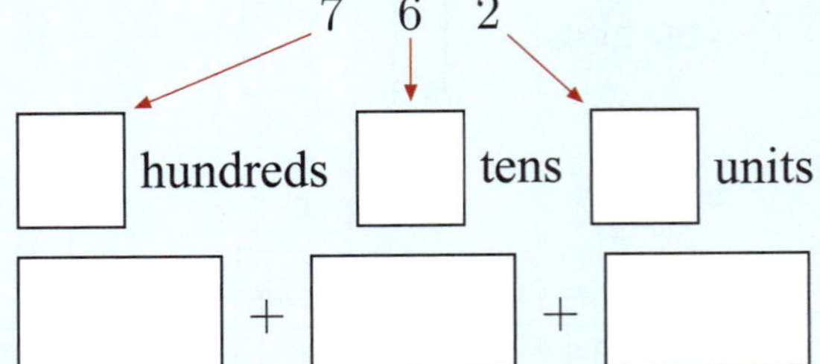

e

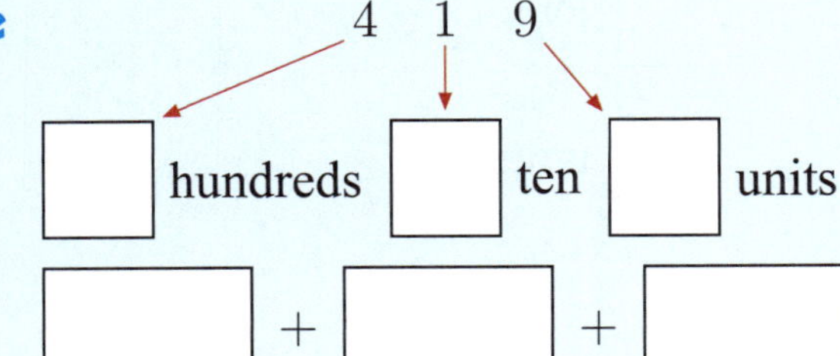

f

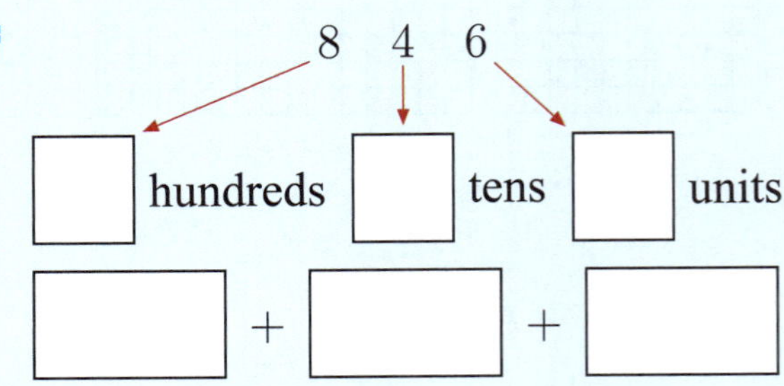

g

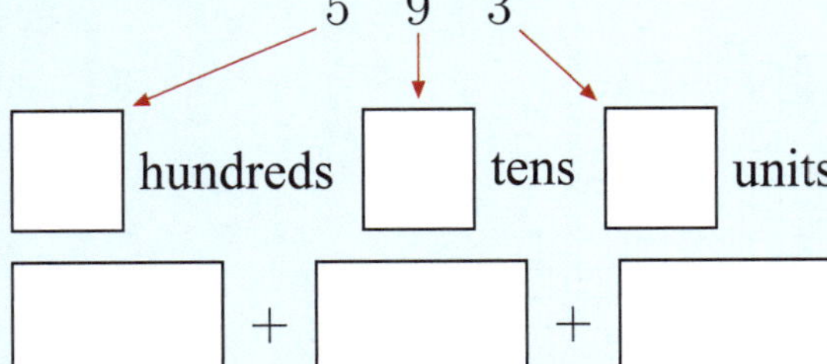

h

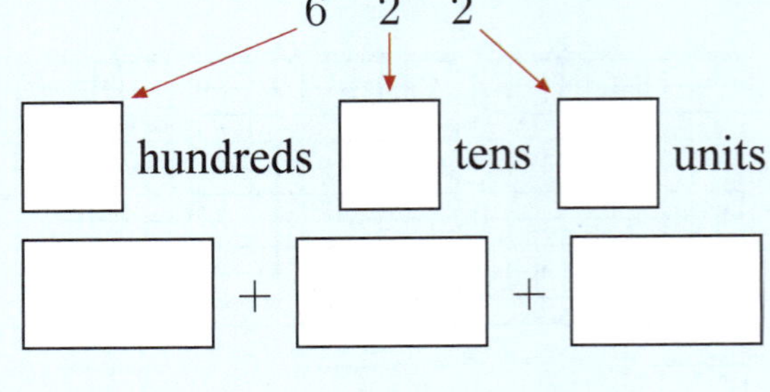

i

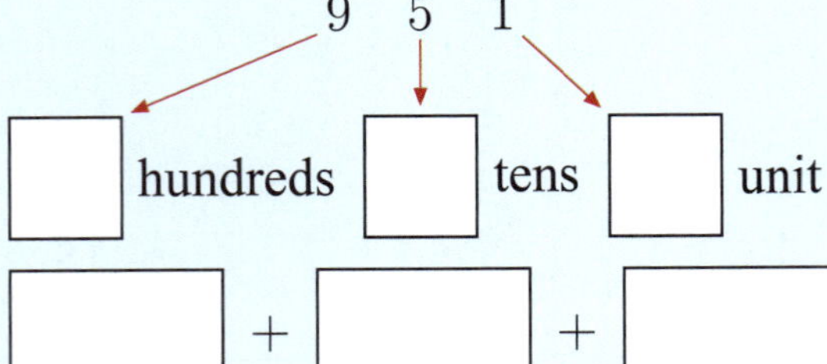

j

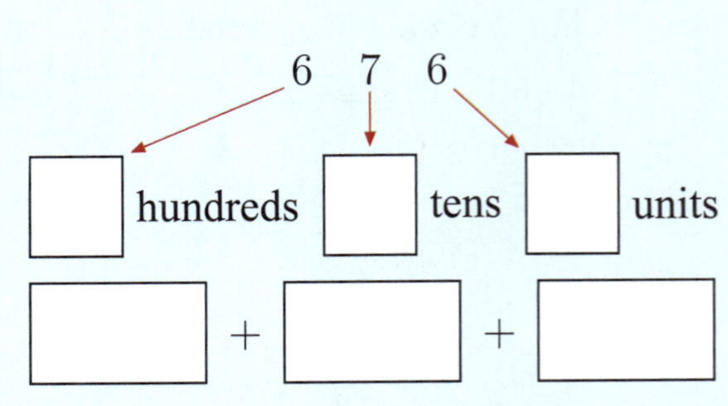

Exercise 13

Write as a numeral:

a one hundred and fifty three ___________

b two hundred and forty nine ___________

c four hundred and thirteen ___________

d six hundred and eighty seven ___________

e nine hundred and sixty one ___________

f eight hundred and twenty four ___________

Exercise 14

Write in words:

a 358 ___________________________________

b 741 ___________________________________

c 489 ___________________________________

d 267 ___________________________________

e 592 ___________________________________

f 845 ___________________________________

Exercise 15

What is the value of the red digit?

In 324, the 2 means 2 tens or 20.

a 618 ___________________ b 542 ___________________

c 317 ___________________ d 493 ___________________

e 356 ___________________ f 782 ___________________

g 950 ___________________ h 806 ___________________

Exercise 16

What number is shown?

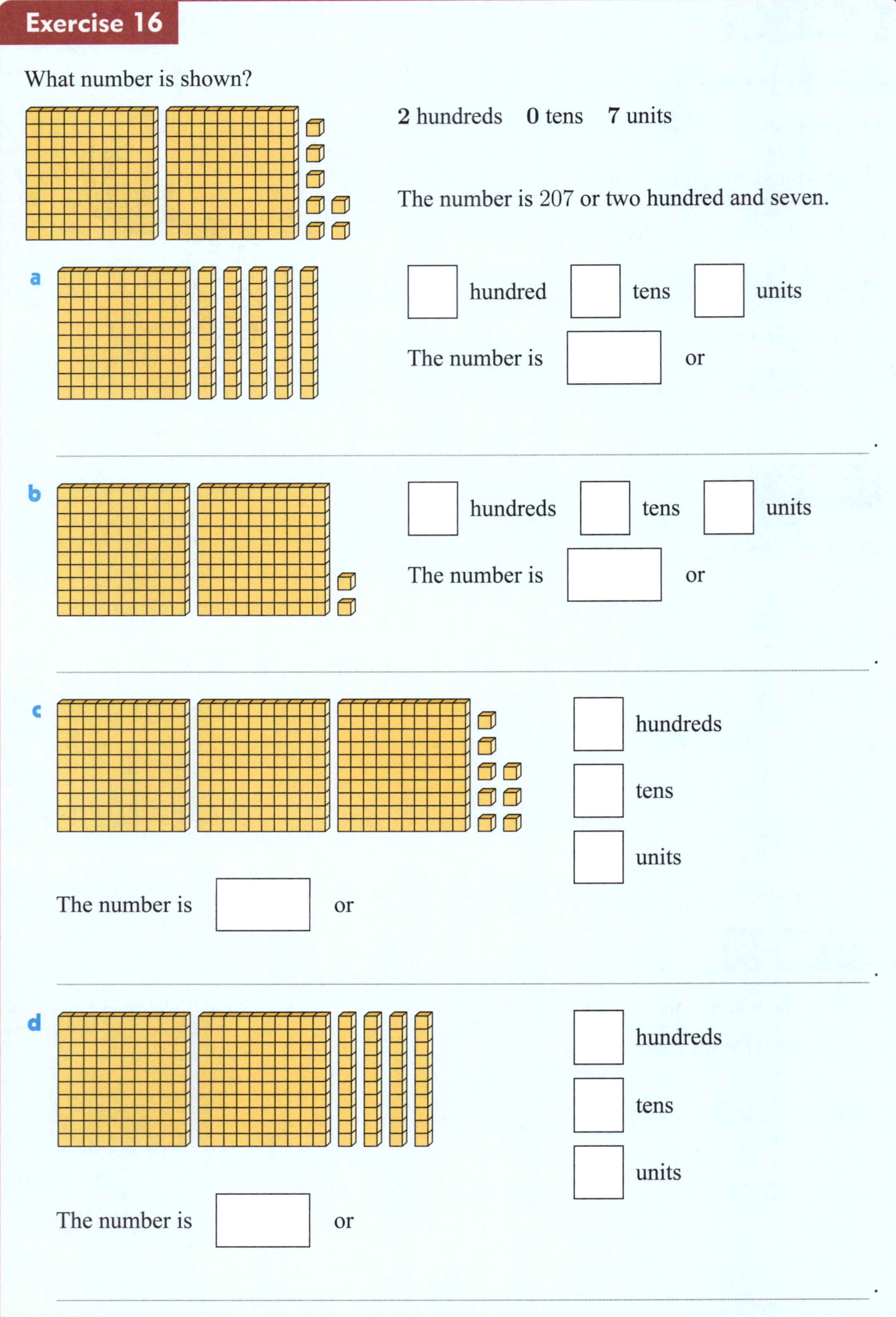

2 hundreds   0 tens   7 units

The number is 207 or two hundred and seven.

a ☐ hundred ☐ tens ☐ units

The number is ☐ or _______________ .

b ☐ hundreds ☐ tens ☐ units

The number is ☐ or _______________ .

c ☐ hundreds

☐ tens

☐ units

The number is ☐ or _______________ .

d ☐ hundreds

☐ tens

☐ units

The number is ☐ or _______________ .

Exercise 17

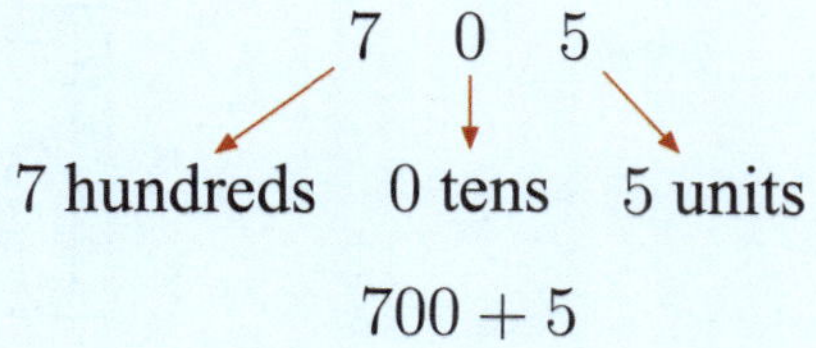

7   0   5

7 hundreds   0 tens   5 units

$700 + 5$

Partition each number using place values:

a

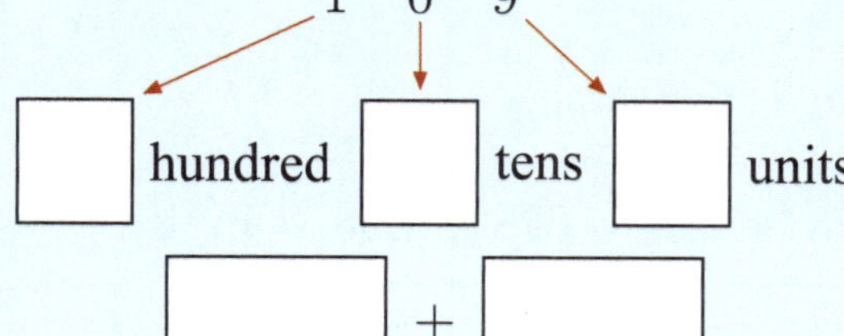

1   0   9

☐ hundred   ☐ tens   ☐ units

☐ + ☐

b

2   4   0

☐ hundreds   ☐ tens   ☐ units

☐ + ☐

c

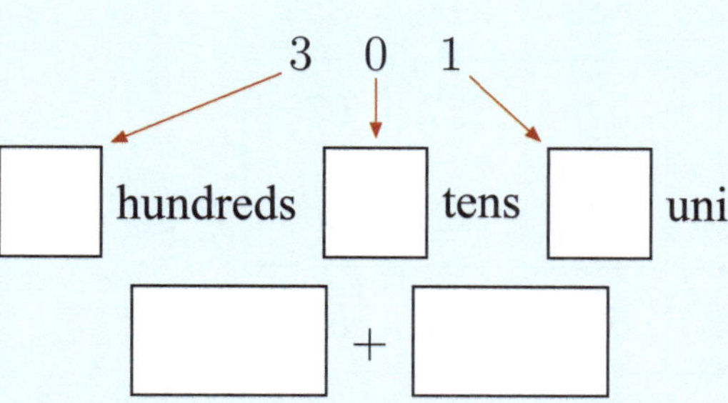

3   0   1

☐ hundreds   ☐ tens   ☐ unit

☐ + ☐

d

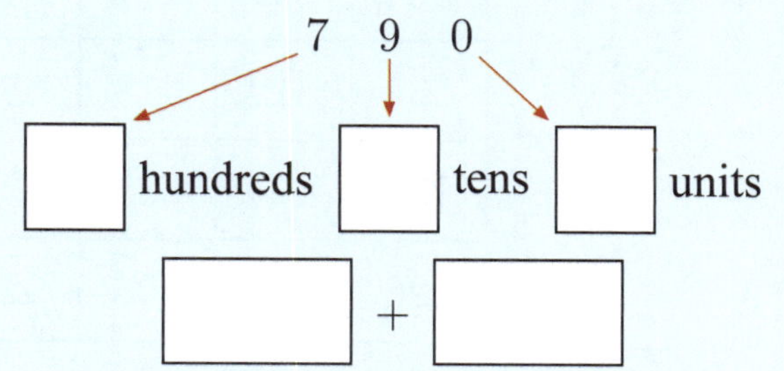

7   9   0

☐ hundreds   ☐ tens   ☐ units

☐ + ☐

Exercise 18

Write in words:

a 206 ___________________________________

b 140 ___________________________________

c 505 ___________________________________

d 780 ___________________________________

e 860 ___________________________________

f 903 ___________________________________

g 407 ___________________________________

h 390 ___________________________________

Activity

Click to practise listening to and reading numbers.

Puzzle

Find your way from the Start to the Finish.

You can move between numbers if any place value is the same.

	637	245	683	323	428	415	628	298	Finish
	930	266	672	177	897	710	736	595	
	964	167	197	579	532	839	216	582	
	534	819	415	680	400	549	237	632	
	137	873	685	624	491	747	968	366	
Start	126	276	306	923	183	387	907	526	

Listening Activity

Click and follow the instructions.

1 _____________   2 _____________   3 _____________   4 _____________

5 _____________   6 _____________   7 _____________   8 _____________

9 _____________   10 _____________

This is an **abacus**. It helps us to see place values.

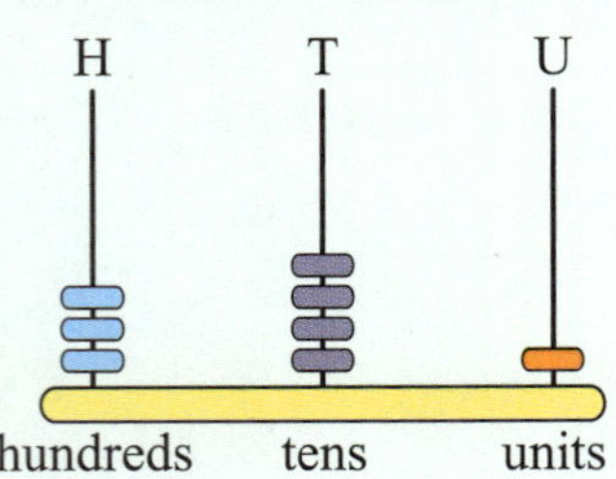

Each column represents a different place value.

3 hundreds + 4 tens + 1 unit = 341

Exercise 19

What number does each abacus show?

a   H   T   U

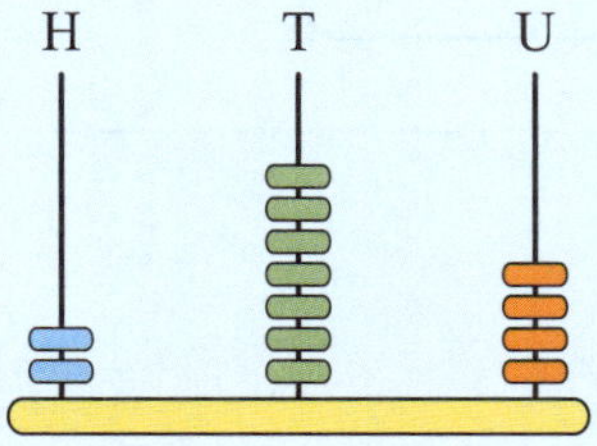

b   H   T   U

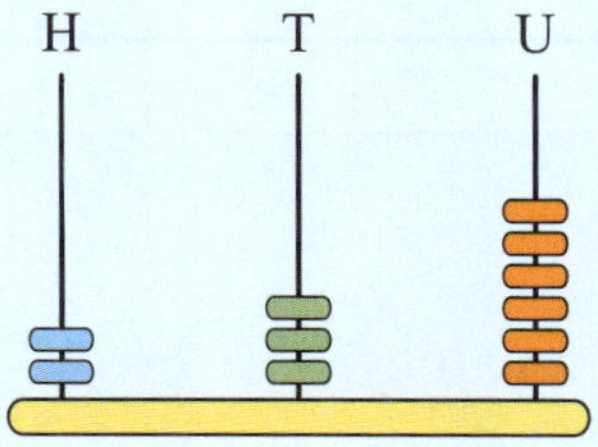

c   H   T   U

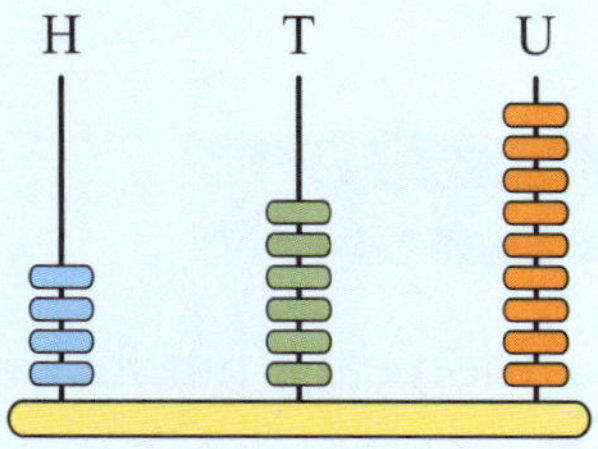

d

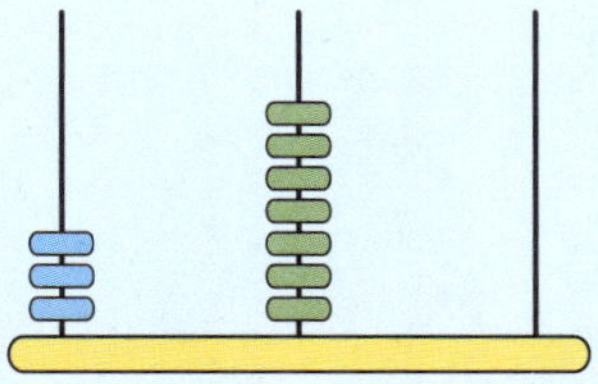

e

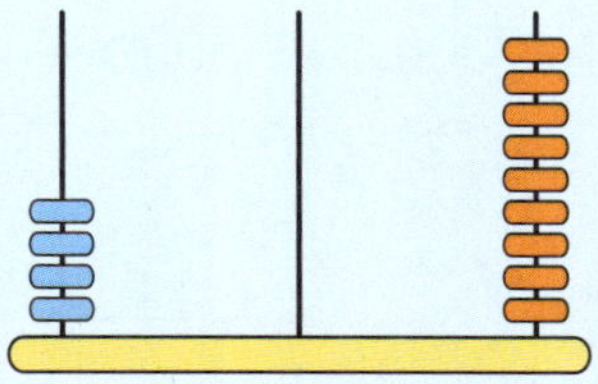

f

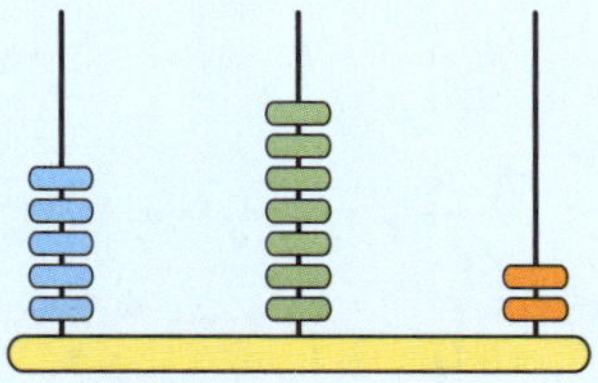

Exercise 20

Show the number on the abacus:

a 376

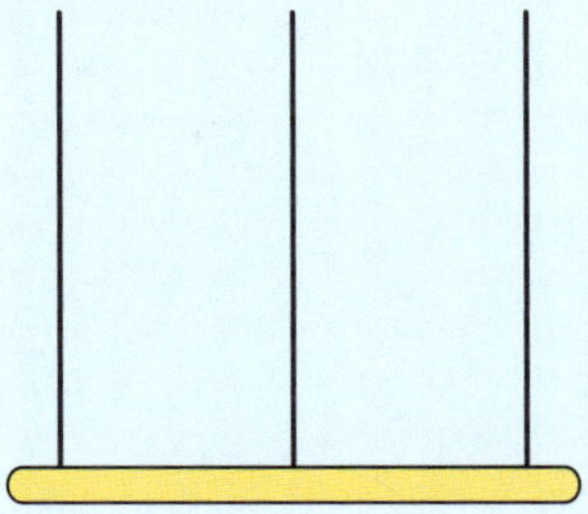

b 500

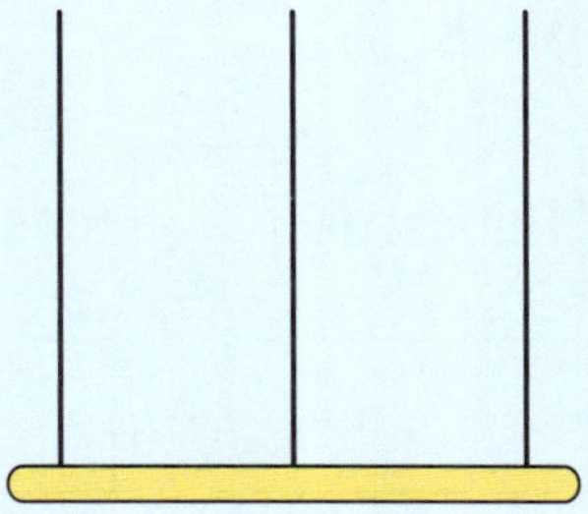

c 851

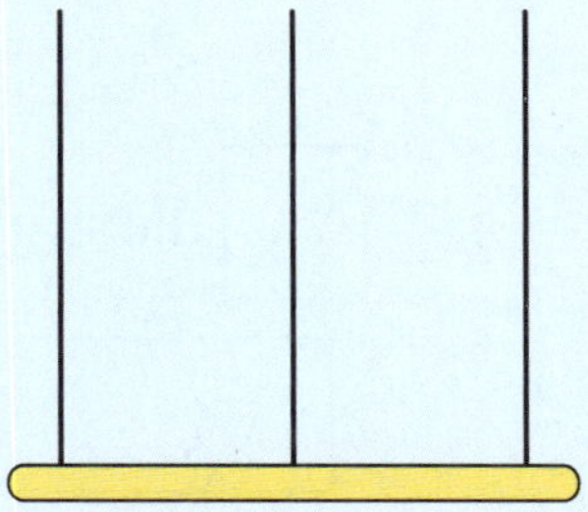

d 904

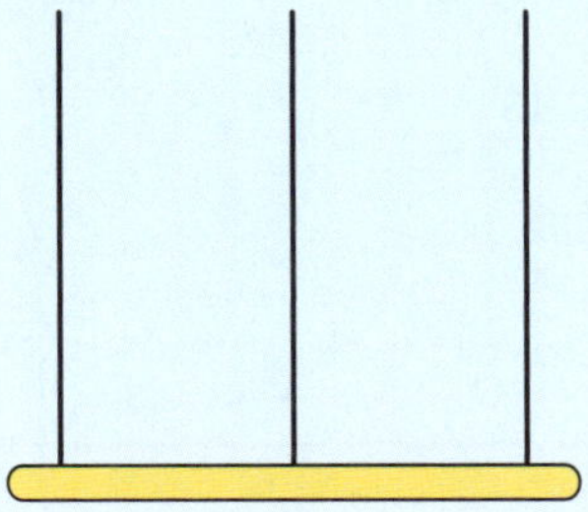

e 429

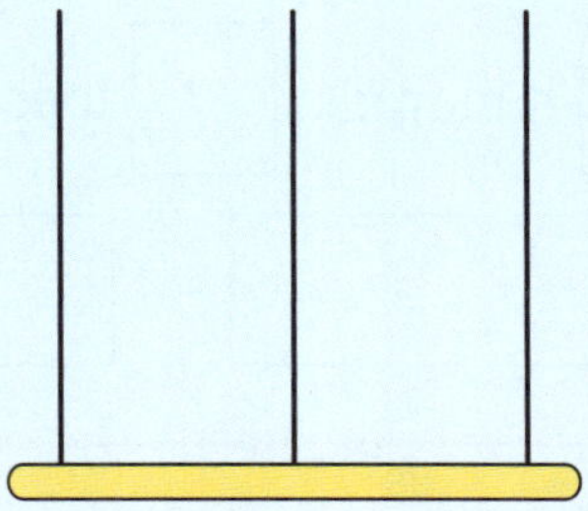

f 280

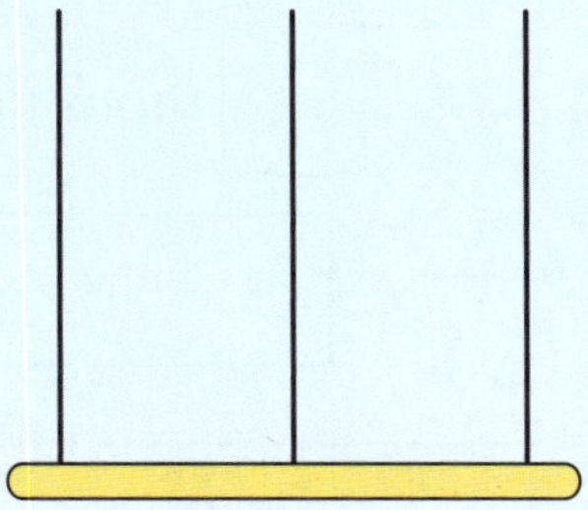

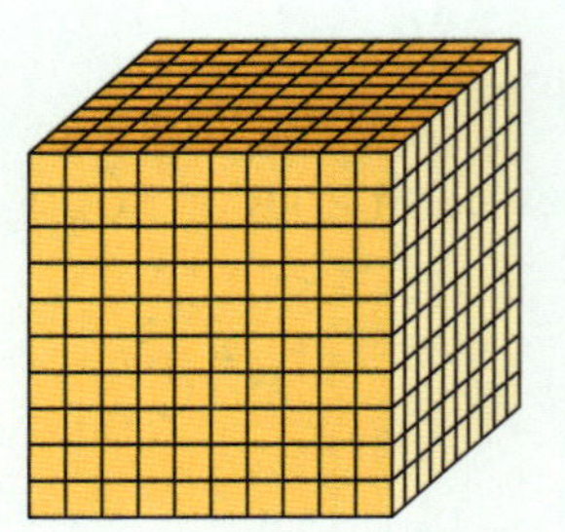

One **thousand** is 10 hundreds

or 1000 units.

Thousands are a place value.

Exercise 21

Partition each number using place values:

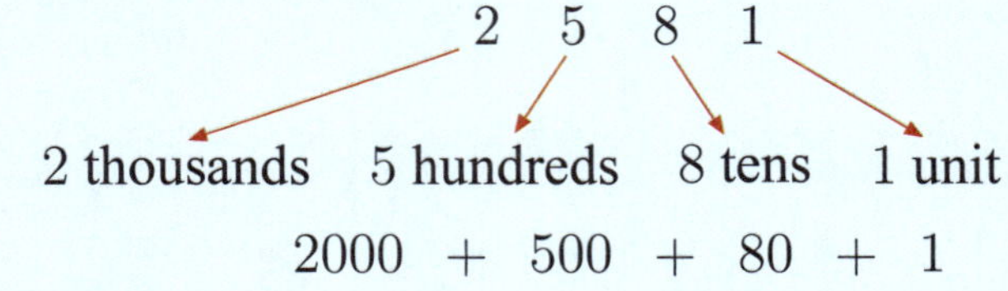

$$2 \quad 5 \quad 8 \quad 1$$

2 thousands   5 hundreds   8 tens   1 unit

$$2000 \ + \ 500 \ + \ 80 \ + \ 1$$

a

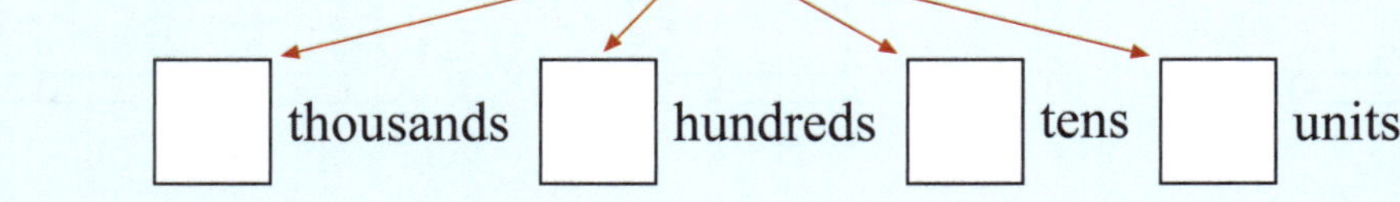

$$3 \quad 6 \quad 9 \quad 4$$

☐ thousands ☐ hundreds ☐ tens ☐ units

☐ + ☐ + ☐ + ☐

b

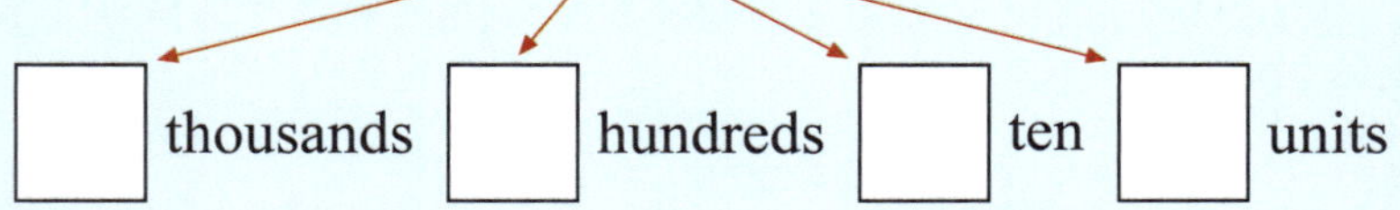

$$6 \quad 7 \quad 1 \quad 5$$

☐ thousands ☐ hundreds ☐ ten ☐ units

☐ + ☐ + ☐ + ☐

c

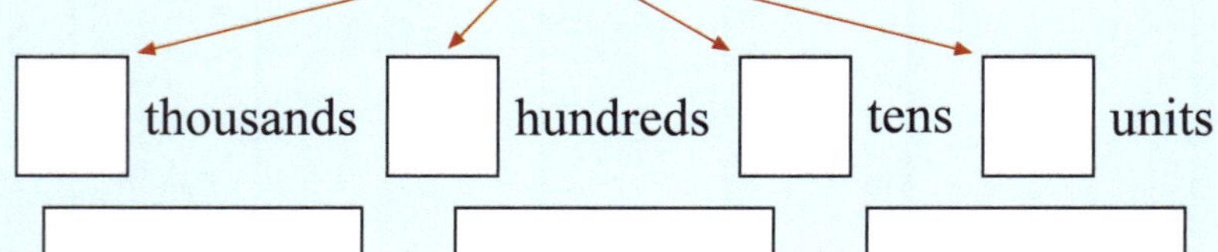

$$5 \quad 0 \quad 8 \quad 2$$

☐ thousands ☐ hundreds ☐ tens ☐ units

☐ + ☐ + ☐

d

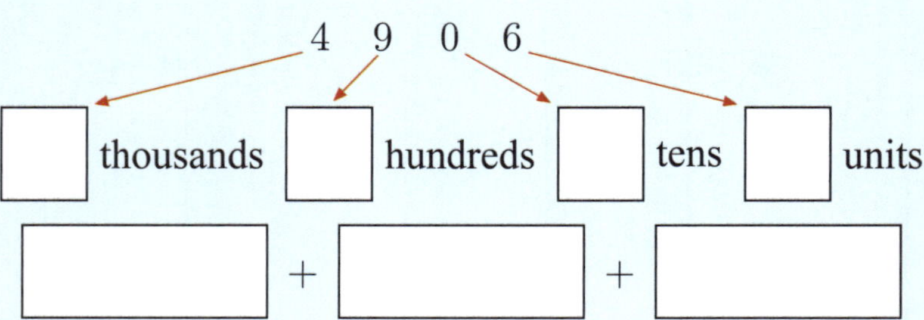

$$4 \quad 9 \quad 0 \quad 6$$

☐ thousands ☐ hundreds ☐ tens ☐ units

☐ + ☐ + ☐

Exercise 22

Practise your spelling: thousand   _______________________________

Exercise 23

numeral   8652

expanded form $8000 + 600 + 50 + 2$

words   eight thousand, six hundred, and fifty two

Write in expanded form and in words:

a 1435 = ________ + ________ + ________ + ________

or ____

b 2147 = ________ + ________ + ________ + ________

or ____

c 4326 = ________ + ________ + ________ + ________

or ____

d 2490 = ________ + ________ + ________

or ____

e 3801 = ________ + ________ + ________

or ____

f 7054 = ________ + ________ + ________

or ____

g 4005 = ________ + ________

or ____

Puzzle

How many *different* words do we need to count from 1 to 9999? _______________

Exercise 24

Show the number on the abacus:

a 2319

b 4721

c 9078

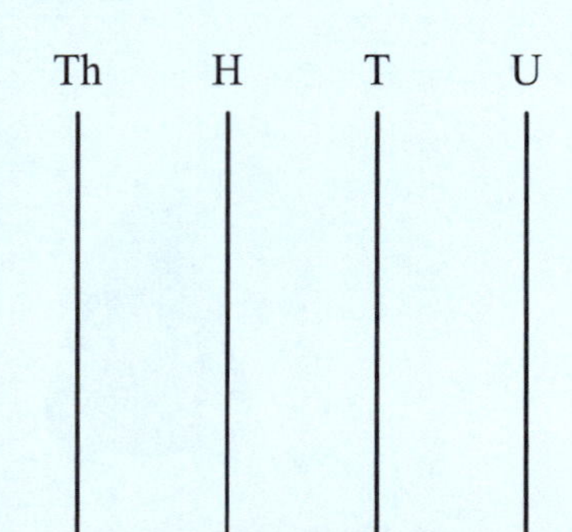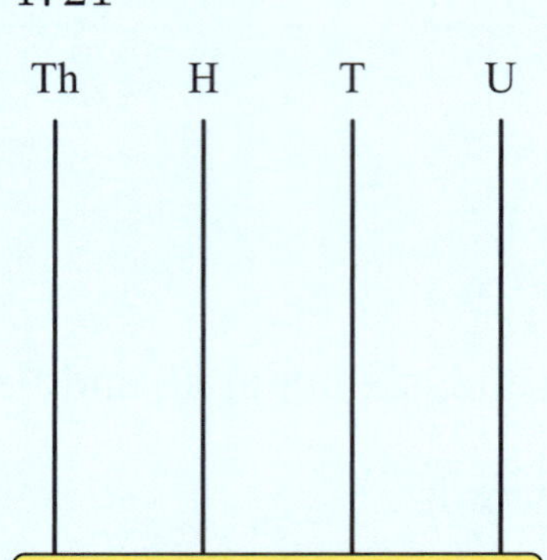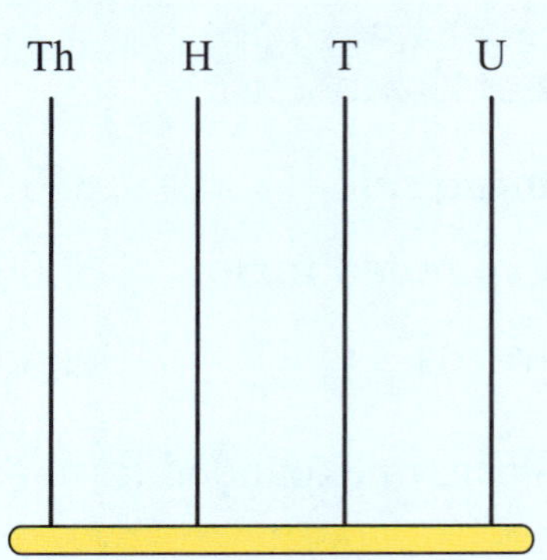

Puzzle

Colour the matching tiles the same:

1234	$2000 + 50$	five thousand, one hundred, and six
5106	$1000 + 200 + 30 + 4$	three thousand and ninety
3009	$2000 + 500 + 90$	one thousand, two hundred, and thirty four
1076	$3000 + 90$	three thousand and nine
2050	$5000 + 100 + 6$	one thousand and seventy six
2590	$3000 + 9$	two thousand and fifty
3090	$1000 + 70 + 6$	two thousand, five hundred, and ninety

Exercise 25

What is the value of the red digit?

In 7213, the 2 means 2 hundreds or 200.

a 1327 _______________________

b 6246 _______________________

c 2361 _______________________

d 1258 _______________________

e 8912 _______________________

f 5129 _______________________

> < means "less than".
>
> > means "greater than".

Exercise 26

4 units or 4

4 is less than 6.

6 units or 6

6 is more than 4.

6 is greater than 4.

$4 < 6$

$6 > 4$

a

5 units or ______

______ is less than ______

______ < ______

9 units or ______

______ is more than ______

______ is greater than ______

______ > ______

b

______ units or ______

______ is more than ______

______ is greater than ______

______ > ______

______ units or ______

______ is less than ______

______ < ______

c

______ units or ______

______ is less than ______

______ < ______

______ units or ______

______ is more than ______

______ is greater than ______

______ > ______

Exercise 27

Complete using $<$ or $>$.

a 5 _____ 7   **b** 7 _____ 4   **c** 3 _____ 2

d 4 _____ 5   **e** 9 _____ 8   **f** 2 _____ 8

g 6 _____ 7   **h** 1 _____ 9   **i** 5 _____ 2

Exercise 28

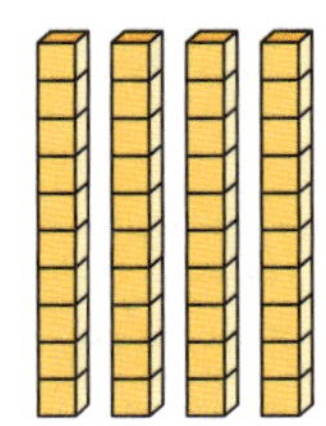

4 tens or 40

40 is less than 70.

$40 < 70$

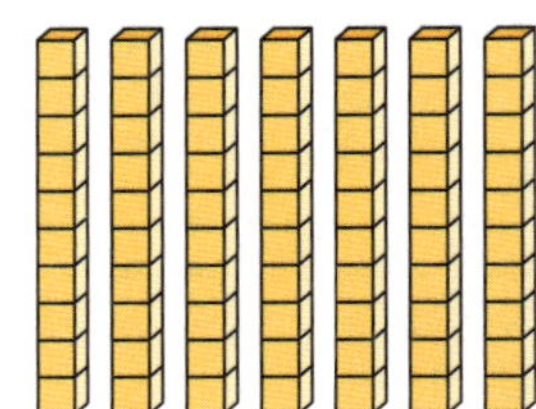

7 tens or 70

70 is greater than 40.

70 is more than 40.

$70 > 40$

a

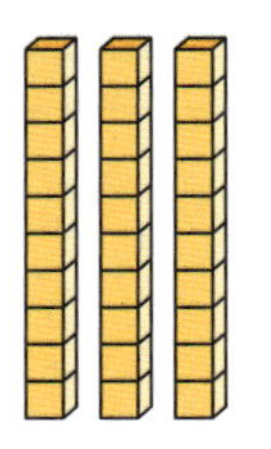

3 tens or _____

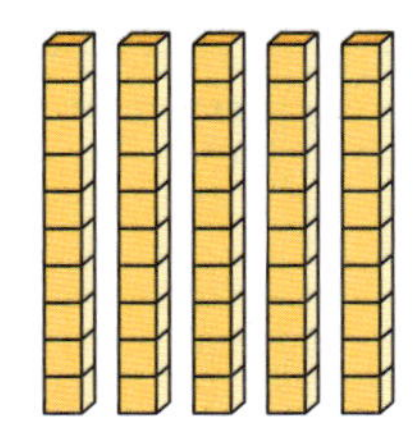

5 tens or _____

_____ is greater than _____

_____ is more than _____

_____ is less than _____

_____ $>$ _____

_____ $<$ _____

b

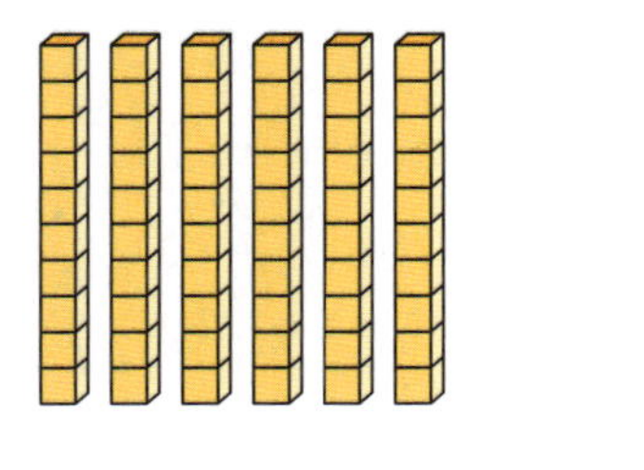

_____ tens or _____

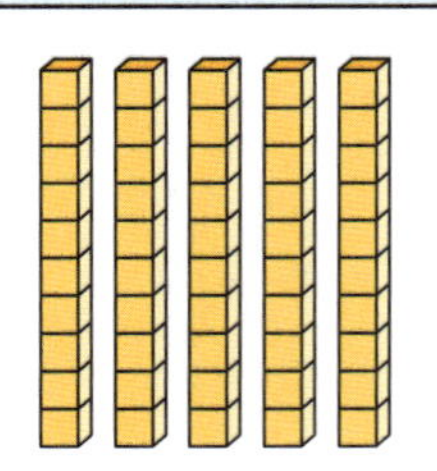

_____ tens or _____

_____ is less than _____

_____ is more than _____

_____ is greater than _____

_____ $<$ _____

_____ $>$ _____

Exercise 29

Complete using $<$ or $>$.

a 30 _____ 40

b 20 _____ 10

c 50 _____ 20

d 60 _____ 80

e 90 _____ 60

f 50 _____ 40

g 20 _____ 70

h 10 _____ 30

i 90 _____ 70

3

5

8

3 is less than 5 and 8.

3 has the least units.

3 is the lowest number.

3 is the least.

8 is more than 3 and 5.

8 has the most units.

8 is the highest number.

8 is the greatest.

Exercise 30

Practise your spelling.

least _______________________

most _______________________

lowest _______________________

highest _______________________

greatest _______________________

Exercise 31

Choose the lowest and highest number from each group.

a 4, 6, 9   lowest _______

highest _______

b 7, 2, 3   lowest _______

highest _______

Exercise 32

Choose the least and greatest number from each group.

a 5, 8, 1   least _______

greatest _______

b 60, 30, 50   least _______

greatest _______

Exercise 33

Circle the number which has the most *units*.

a 3   7   5

b 16   21   8

c 34   28   16

d 9   27   53

Exercise 34

Circle the number which has the least *tens*.

a 90   48   74

b 68   17   88

c 146   392   650

d 575   607   924

Exercise 35

258   814   367   546

Write the number with the most:

a units __________

b tens __________

c hundreds __________

To *compare* numbers, look at the largest place value first.

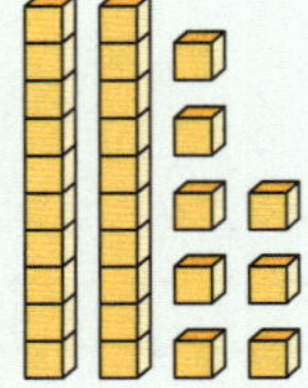

28 has 2 tens
and 8 units.

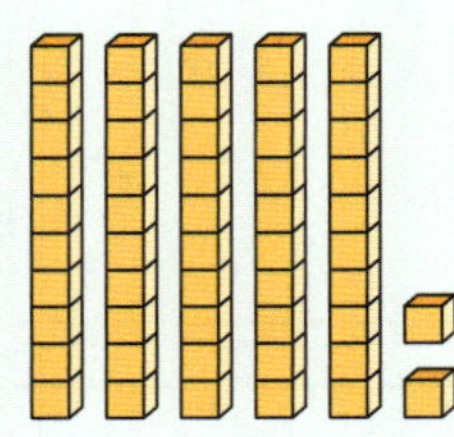

52 has 5 tens
and 2 units.

52 has more *tens* than 28
so 52 is *greater than* 28.

Exercise 36

Complete using < or >.

a 15 _____ 27

b 21 _____ 17

c 35 _____ 40

d 45 _____ 16

e 91 _____ 65

f 42 _____ 73

g 21 _____ 37

h 46 _____ 37

i 53 _____ 94

To arrange numbers in order, look at the largest place value first.

Ascending order means from lowest to highest.

Descending order means from highest to lowest.

Exercise 37

Practise your spelling.

ascending ___________________________   ___________________________

descending ___________________________   ___________________________

Exercise 38

27   58   41
↑   ↑   ↑
2 tens   5 tens   4 tens

In ascending order, we write 27, 41, 58.

Arrange the numbers in ascending order.

a 5   7   4   ___________________

b 28   17   36   ___________________

c 47   33   62   ___________________

d 25   9   18   ___________________

e 635 864 239 ___________________

f 126 87   353 ___________________

Exercise 39

38   12   43
↑   ↑   ↑
3 tens   1 ten   4 tens

In descending order, we write 43, 38, 12.

Arrange the numbers in descending order.

a 8   4   9   ___________________

b 27   16   39   ___________________

c 61   27   38   ___________________

d 46   7   15   ___________________

e 167 430 812 ___________________

f 101 85   229 ___________________

> A **cardinal number** is used for **counting**.
>
> 1, 2, 3, and 4 are cardinal numbers.
>
> An **ordinal number** is used for **order**.
>
> The positions first, second, third, and fourth are ordinal numbers.

Listening Activity

This Listening Activity will help you learn and say the words for numbers that describe order.

Exercise 40

Copy each ordinal number and learn its spelling.

1st	first	__________
2nd	second	__________
3rd	third	__________
4th	fourth	__________
5th	fifth	__________
6th	sixth	__________
7th	seventh	__________
8th	eighth	__________
9th	ninth	__________
10th	tenth	__________
11th	eleventh	__________
12th	twelfth	__________
13th	thirteenth	__________
14th	fourteenth	__________
15th	fifteenth	__________

16th	sixteenth	
17th	seventeenth	
18th	eighteenth	
19th	nineteenth	
20th	twentieth	
21st	twenty first	
22nd	twenty second	
23rd	twenty third	
24th	twenty fourth	
25th	twenty fifth	
26th	twenty sixth	
27th	twenty seventh	
28th	twenty eighth	
29th	twenty ninth	
30th	thirtieth	
31st	thirty first	

Activity

a the sixth letter of copperhead

b the fifth letter of taipan

c the ninth letter of boomslang

d the second letter of cobra

e the fourth letter of viper

f the seventh letter of anaconda

Activity

Label the animals using these clues:

Maud the cat is first in line.

The 4th animal is Lizzy.

The 8th animal is Betty.

The twelfth animal is Bert.

The fifth animal is Marg.

The 1st zebra is Kath.

The third cat is George.

The second cat is Colin.

The third zebra is Mervyn.

The second chicken is Len.

The ninth animal is Maureen.

The tenth animal is Keith.

Maud

Listening Activity

Click and listen to the instructions.

1 ___________________ 2 ___________________ 3 ___________________

4 ____

5 _________ , _________ , _________

Challenge

Lucy has 9 beads. She places all of them on the abacus.

1 Draw the lowest number she can make.

2 Draw the highest number she can make.

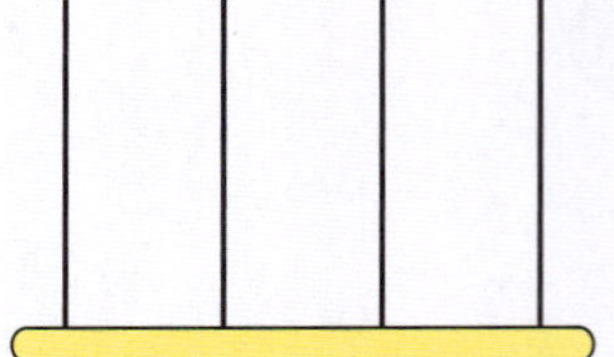

3 Draw the highest number she can make with at least one bead in each column.

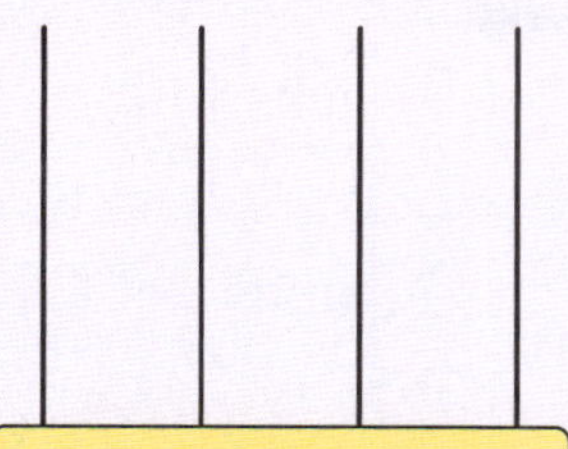

4 Draw the lowest number she can make with at least one bead in each column.

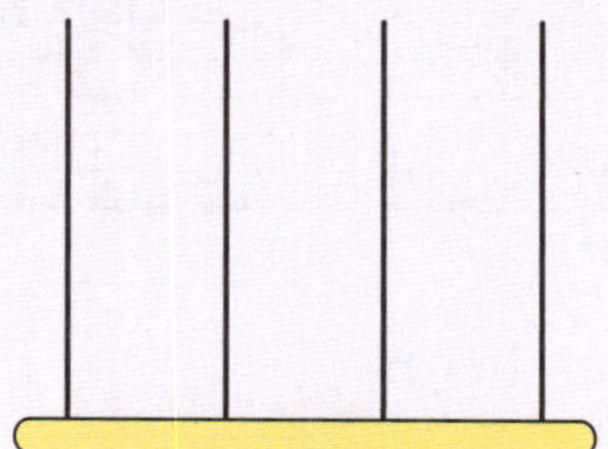

5 Draw a number greater than 1000 she can make with digits in descending order.

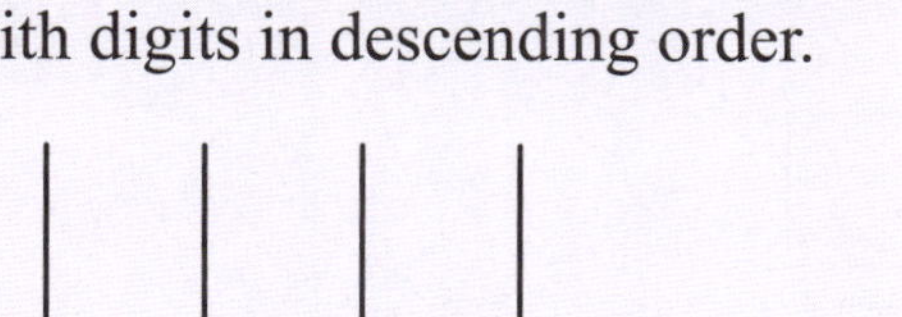

Revision

1 What numbers are shown?

a

☐ tens and ☐ units

☐ + ☐ = ☐

b

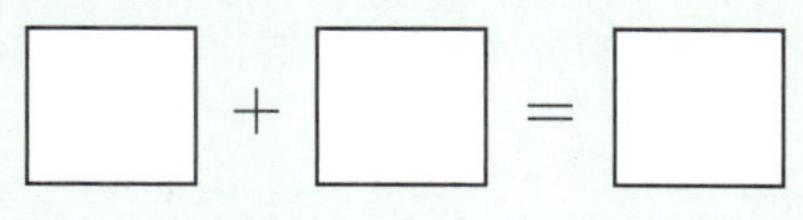

☐ tens and ☐ units

☐ + ☐ = ☐

c

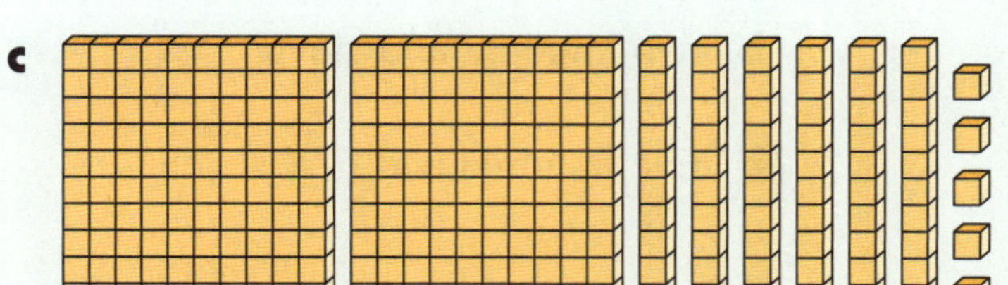

☐ hundreds

☐ tens

☐ units

The number is ☐ or

d 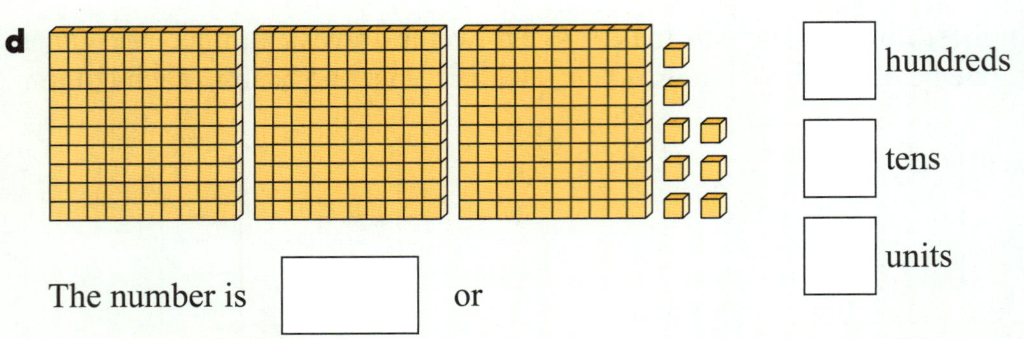

☐ hundreds

☐ tens

☐ units

The number is ☐ or

---

2 Write as a numeral:

a 6 hundreds, 8 tens, and 3 units is ☐

b 9 hundreds, 1 ten, and 7 units is ☐

c 7 hundreds and 2 tens is ☐

3 Partition each number using place values:

a 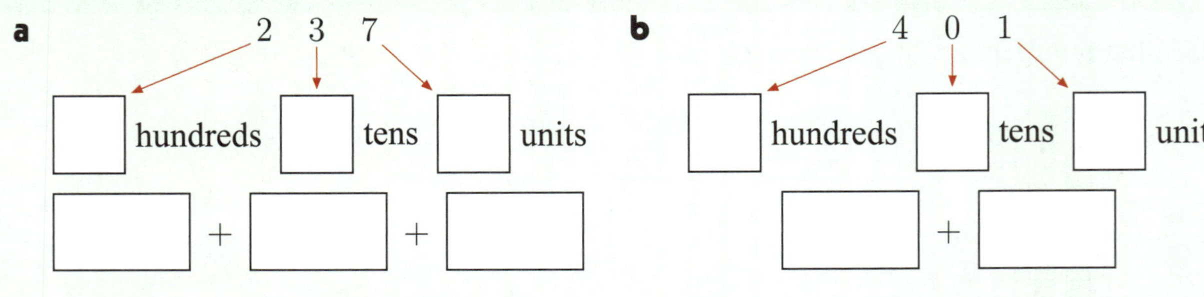

2 3 7

☐ hundreds ☐ tens ☐ units

☐ + ☐ + ☐

b 4 0 1

☐ hundreds ☐ tens ☐ unit

☐ + ☐

c 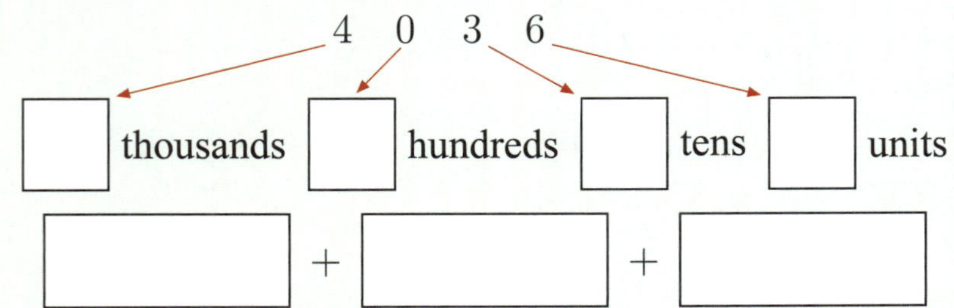

4 0 3 6

☐ thousands ☐ hundreds ☐ tens ☐ units

☐ + ☐ + ☐

4 Write as a numeral:

a eighty five _______

b two hundred and eleven _______

c six hundred and ninety _______

d five hundred and fifty seven _______

5 What is the value of the red digit?

a 305 _______________________

b 742 _______________________

c 869 _______________________

d 936 _______________________

e 4871 _______________________

f 2583 _______________________

6 Write in words:

a 78

b 109

c 340

d 862

7 Write in expanded form and in words:

a 2306 = _______ + _______ + _______

or _______

b 8074 = _______ + _______ + _______

or _______

8 What number does each abacus show?

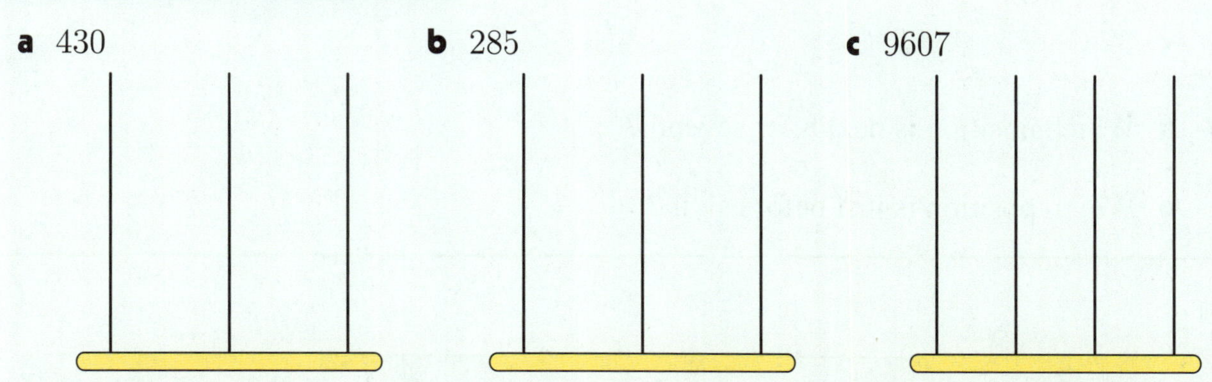

a   **b**   **c**

9 Show the number on the abacus.

a 430   **b** 285   **c** 9607

10 Complete using < or >.

a 4 ___ 3   **b** 1 ___ 6   **c** 7 ___ 9

d 70 ___ 40   **e** 20 ___ 50   **f** 80 ___ 60

11 Write down the lowest and highest number.

8, 5, 6   lowest = _______   highest = _______

12 Write down the least and greatest number.

     2, 1, 7, 4   least = _______   greatest = _______

13 **a** Circle the number which has the least *units*:

         25   37   62   56

  b Circle the number which has the most *tens*:

         135   91   77   48

  c Circle the number which has the least *hundreds*:

         623   417   925   386

14 Complete using $<$ or $>$.

  a 46 _____ 70   **b** 82 _____ 98   **c** 67 _____ 14

15 Arrange the numbers from lowest to highest.

  a 32   29   51 ________________________________

  b 109   68   97 ________________________________

  c 325   419   248 ________________________________

16 Arrange the numbers from highest to lowest.

  a 75   57   82 ________________________________

  b 88   79   100 ________________________________

  c 325   685   490 ________________________________

17 **a** Which position is next after seventh? ________________________________

  b Which position is just before sixth? ________________________________

Discussion

Here are some words we can use to describe the **position** or **location** of an object.

With your class, think of a sentence which uses each word or words.

to the left of	above	in front of	between
to the right of	over	behind	centre
alongside	below	outside	middle
next to	under	inside	
	underneath		near
	beneath		far

Exercise 1

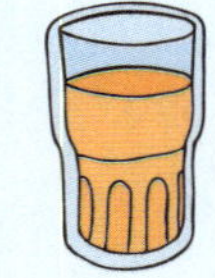
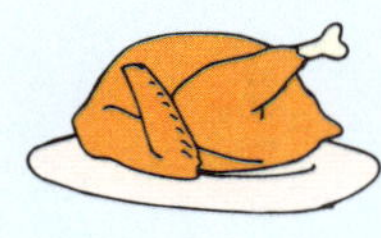

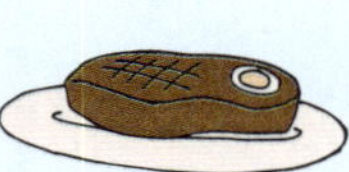

Complete each sentence using *left*, *right*, *between*, *above*, *below*, or *middle*:

a The chicken is _______________________ the cake.

b The salad is _______________________ the chicken and the steak.

c The fries are _______________________ the ice cream.

d The spaghetti is to the _______________________ of the fries.

e The ice cream is in the _______________________ .

f The burger is to the _______________________ of the fries.

Name *two* items under the burger. _______________________ _______________________

Discussion

Compare "over", "above", and "on top of".

Do these words always mean the same thing?

Discuss which words you would choose to complete:

- The bird is _________________ the building.

- The antenna is _________________ the building.

Exercise 2

Complete each sentence using *beneath*, *between*, *inside*, *outside*, or *on top of*.

a The snowman is _________________ .

b The chair is _________________ the table and the window.

c The puppy is _________________ the table.

d The bird is _________________ the cage _________________ the table.

Puzzle

Draw a cross ✗ on the tile that has a red tile above it, a blue tile to its left, and is below a green tile.

Exercise 3

Complete each sentence describing the position of the items on Pippa's bookshelf:

a The purple hat is

__________ __________ __________

of the suitcase.

b The photographs are

__________________ the purple hat.

c There are two dark blue books

__________________ the black hat and the globe.

d The giraffe is

__________ __________ __________

of the zebra.

e The black hat is ______________ the three brown books.

f The suitcase is ____

____ .

g The orange maths book is ____________________________

____ .

h The globe is ____________________________________

____ .

i The book about China is ________________________

____ .

Activity

Draw a picture to match this description:

On the left side, water flows down rocks into the pond.

Behind the pond are three tall trees.

A lily pad is in the middle of the pond. A frog sits on top of it.

Alongside the lily pad are eight ducklings. Their mother is on the right.

Activity

Use this picture to write 5 longer sentences describing the position of objects.

For example:

> The spider is to the left of the window with the open green curtains.

1 _______________________________

____

____

____

2 _______________________________

____

____

____

3 _______________________________

____

____

____

4 _______________________________

____

____

5 _______________________________

____

Exercise 4

This is a **plan** of Mrs Nguyen's classroom.

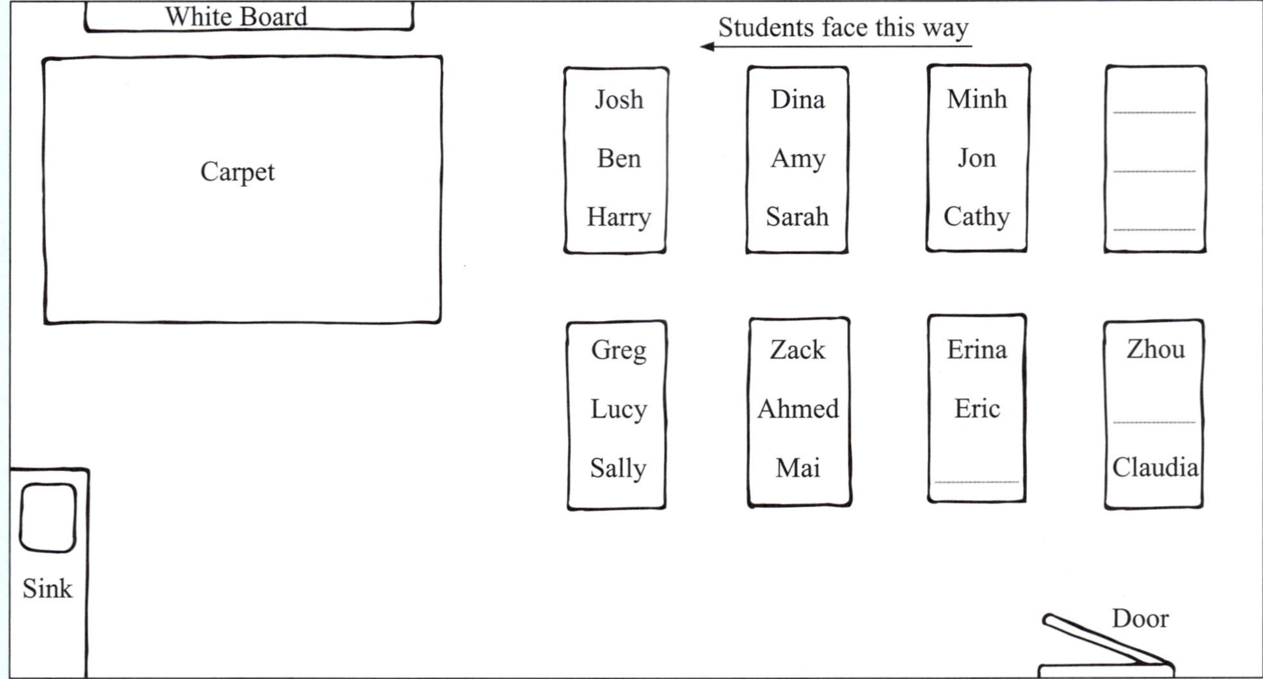

a Who sits between Josh and Harry? _______________________

b Who sits in front of Ahmed? _______________________

c George sits next to Claudia. Mark George's seat on the plan.

d Who sits on Amy's right hand side? _______________________

e Jin sits behind Mai. Mark Jin's seat on the plan.

f Rachel sits behind Minh. Mark Rachel's seat on the plan.

g Hannah sits between Alex and Rachel. Mark Hannah's seat and Alex's seat on the plan.

h Draw a bookcase between the sink and the door.

i Mrs Nguyen's desk is just off the carpet, and she likes to face her class.
Draw Mrs Nguyen's desk. Draw an arrow to show the direction she faces.

Discussion

Look around your own classroom.

Make a list of items that Mrs Nguyen may have left off her plan.

Activity

In this Activity you will draw a plan of your classroom.

You will need: pencil, ruler

What to do:

1 Use a ruler to carefully draw the shape of your classroom in the space below.

2 Mark the door.

3 Begin to draw the desks and other large objects. Press lightly until you are sure you have included everything correctly, then draw them using a ruler.

4 Mark where you and 5 other classmates sit in the classroom.

Discussion

Two objects or places are *opposite* one another if they are either side of a space.

The Fish and Chip shop is *opposite* the post box. The road is between them.

Discuss some sentences which use the word "opposite".

Exercise 5

A helicopter can move in all directions.

Complete using *forwards*, *backwards*, *left*, *right*, *up*, and *down*:

Sideways can mean to the _______________ or to the _______________ .

A **turn** is a change of direction.

If we continue in the *same* direction, we are going **straight ahead**.

Activity

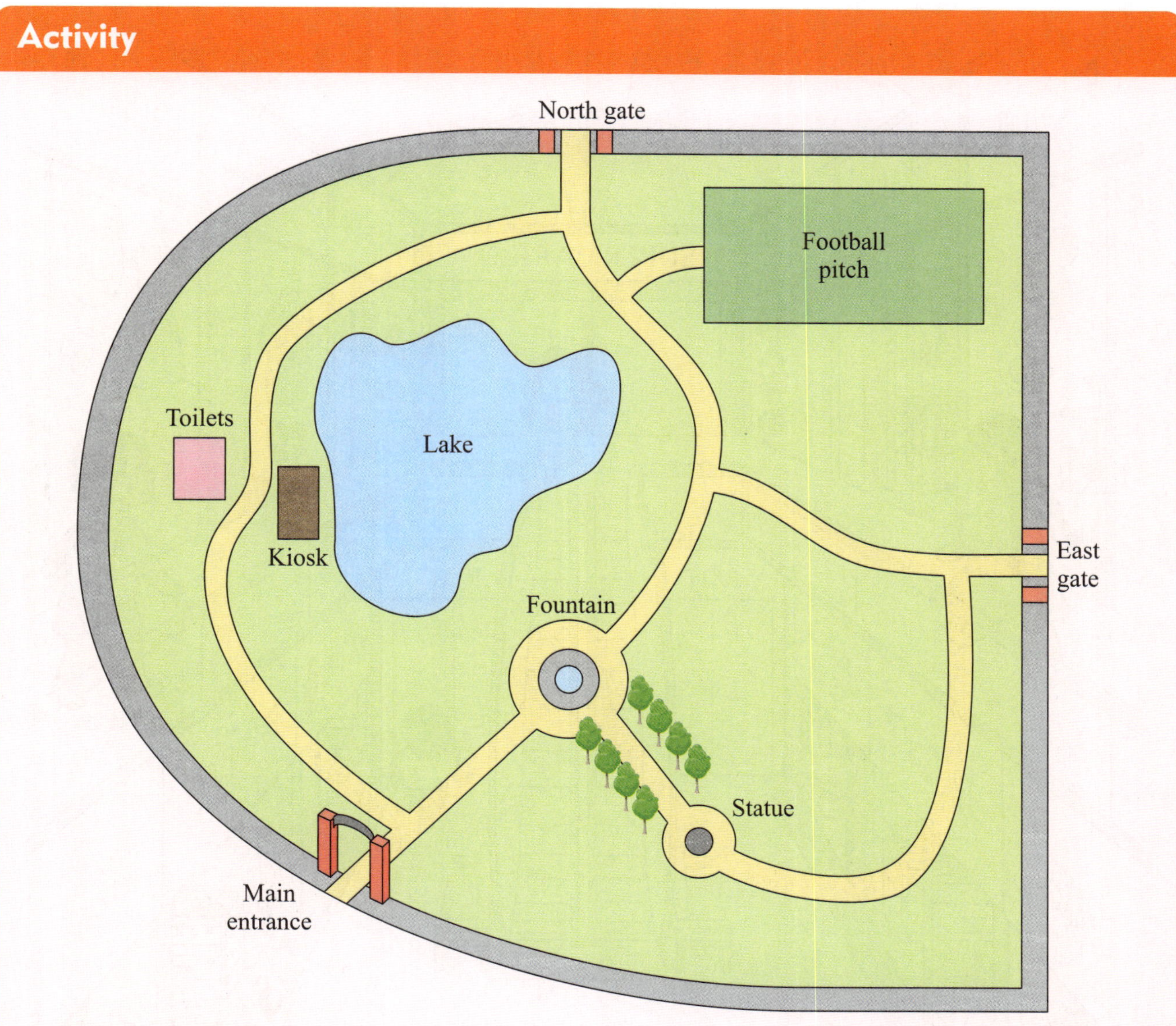

1   Draw the path followed by each person.

   a   Gabriela enters at the East gate. She turns left and walks to the statue, then continues on the path between the trees to the fountain.

   b   Liliana enters at the North gate. She turns right and walks around the lake to the kiosk, where she buys an ice cream.

2   Describe how Liliana can get to the fountain, to meet Gabriela.

Exercise 6

Celine wants to get to the bookshop. She needs to go *up* the escalator, then turn *left*. The bookshop is *between* the pet store and the sporting goods.

Describe how:

a Tam can get to the shoe store

b Ben can get to the pet store

c David can get to the supermarket

d Ruby can get to the optician

e Maggie can find a seat to sit on.

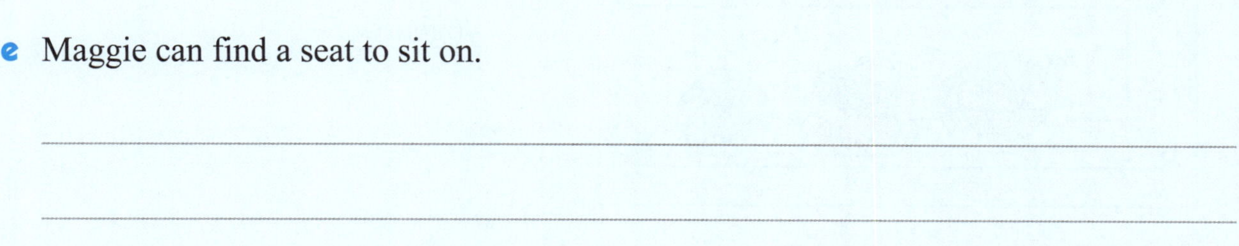

Revision

1 Match the words with opposite meanings.

inside   underneath

near   right

behind   far

outside   in front of

above   left

2

Complete each sentence describing the position of the items in Jun's refrigerator:

a The cheese is

_____________ _____________ _____________

of the cream.

b The eggs are _____________ the milk and the juice.

c The carrots are _____________ the mushrooms.

d The juice is _____________ the yoghurt.

e The lettuce is _____________ _____________ _____________ of the tomatoes.

f The cream is _____________________________________

_____________________________________ .

Discussion

What words tell us that we need to add?

___________________   ___________________   ___________________

___________________   ___________________

To find $7 + 8$, we start at 7 then add 8.

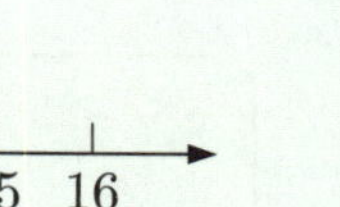

$7 + 8 = 15$

Exercise 1

Use the number line to help you add:

a
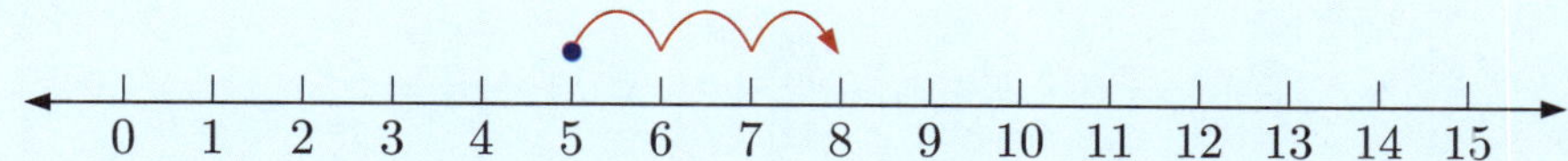

$5 + 3 = $ _______

b

$2 + 7 = $ _______

c

$6 + 4 = $ _______

d

$4 + 7 = $ _______

e

$9 + 3 = $ _______

f

$5 + 9 = $ _______

Exercise 2

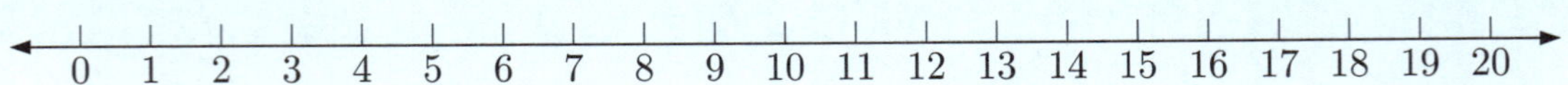

Write down the result of each addition:

a $6 + 9 =$ _____

b $8 + 8 =$ _____

c 12 add 6 = _____

d 9 plus 7 = _____

e 5 add 13 = _____

f 7 plus 11 = _____

Mental additions are done without a
pencil and paper, or apparatus.

$$5 \quad + \quad 3 \quad = \quad 8$$

Exercise 3

Write down the result of each addition:

a $3 + 2 =$ _____

b $3 + 3 =$ _____

c $4 + 3 =$ _____

d 4 add 2 = _____

e 5 plus 2 = _____

f 4 add 6 = _____

g $5 + 5 =$ _____

h 7 plus 3 = _____

i $8 + 2 =$ _____

j $8 + 6 =$ _____

k 7 add 9 = _____

l 9 plus 6 = _____

m 4 plus 8 = _____

n $5 + 7 =$ _____

o 8 add 9 = _____

Exercise 4

a Add together 6 and 3. _____

b Find the total of 6 and 4. _____

c Find the sum of 4 and 4. _____

d The total of 2 and 6 is _____ .

e Add together 5 and 6. _____

f Find the sum of 6 and 6. _____

g The number 4 more than 9 is _____ .

Puzzle

1   2   3   4   5   6

If I add any two of these numbers, the possible totals I can make are:

______, ______, ______, ______, ______, ______, ______, ______, and ______ .

There are [2] ways to make the total [3] . They are: $1 + 2$ or $2 + 1$.

There are [ ] ways to make the total [ ] .

They are: ____ .

There are [ ] ways to make the total [ ] .

They are: ____ .

There are [ ] ways to make the total [ ] .

They are: ____ .

There are [ ] ways to make the total [ ] .

They are: ____ .

There are [ ] ways to make the total [ ] .

They are: ____ .

There are [ ] ways to make the total [ ] .

They are: ____ .

There are [ ] ways to make the total [ ] .

They are: ____ .

There are [ ] ways to make the total [ ] .

They are: ____ .

Exercise 5

The number 0 is called ________________ .

Complete each pair of additions:

a $9 + 0 =$ _______

$0 + 9 =$ _______

b $0 + 12 =$ _______

$12 + 0 =$ _______

c $38 + 0 =$ _______

$0 + 38 =$ _______

Exercise 6

Complete these additions.

$0 + 2 =$ _______

$10 + 2 =$ _______

$20 + 2 =$ _______

$30 + 2 =$ _______

$40 + 2 =$ _______

$0 + 4 =$ _______

$10 + 4 =$ _______

$30 + 4 =$ _______

$50 + 4 =$ _______

$70 + 4 =$ _______

$80 + 9 =$ _______

$60 + 9 =$ _______

$40 + 9 =$ _______

$20 + 9 =$ _______

$0 + 9 =$ _______

Exercise 7

Write pairs of numbers that total 13.

$13 =$ ☐ $+$ ☐ $13 =$ ☐ $+$ ☐ $13 =$ ☐ $+$ ☐

$13 =$ ☐ $+$ ☐ $13 =$ ☐ $+$ ☐ $13 =$ ☐ $+$ ☐

$13 =$ ☐ $+$ ☐ $13 =$ ☐ $+$ ☐ $13 =$ ☐ $+$ ☐

$13 =$ ☐ $+$ ☐ $13 =$ ☐ $+$ ☐ $13 =$ ☐ $+$ ☐

$13 =$ ☐ $+$ ☐ $13 =$ ☐ $+$ ☐

Exercise 8

Complete these pairs of additions.

a
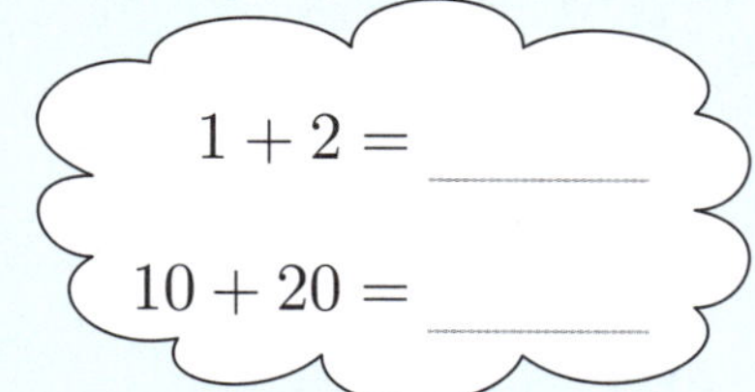

b

c

d

Discussion

Describe the pattern in the pairs of additions above.

Exercise 9

Complete these additions.

a $10 + 40 = $ _______

b $40 + 30 = $ _______

c $50 + 10 = $ _______

d $20 + 40 = $ _______

e $60 + 30 = $ _______

f $20 + 80 = $ _______

g $40 + 60 = $ _______

h $70 + 50 = $ _______

i $50 + 60 = $ _______

j $90 + 70 = $ _______

k $70 + 80 = $ _______

l $80 + 90 = $ _______

Exercise 10

Complete these additions.

a $5 + $ _______ $= 11$

b $7 + $ _______ $= 12$

c _______ $+ 9 = 14$

d _______ $+ 8 = 13$

e $6 + $ _______ $= 16$

f _______ $+ 3 = 23$

g $9 + $ _______ $= 18$

h _______ $+ 5 = 45$

i $20 + $ _______ $= 70$

j $30 + $ _______ $= 36$

k _______ $+ 6 = 15$

l _______ $+ 80 = 130$

Exercise 11

Write the result of each addition:

a $2 + 1 + 3 =$ _______

b $3 + 2 + 1 =$ _______

c $2 + 2 + 4 =$ _______

d $4 + 1 + 5 =$ _______

e $3 + 2 + 6 =$ _______

f $5 + 5 + 4 =$ _______

g $6 + 4 + 3 =$ _______

h $7 + 3 + 6 =$ _______

i $8 + 2 + 7 =$ _______

Game

Click the icon to practise your addition.

Exercise 12

Write an addition to solve each problem:

a Jacinta has 3 strawberries and 8 raspberries.

In total, Jacinta has _______ + _______ = _______ berries.

b Aslan plants 40 orange trees and 30 apple trees in his orchard.

In total, Aslan has planted _______ + _______ = _______ trees.

c Brianna has 5 skirts, 4 dresses, and 6 pairs of jeans.

Altogether, Brianna has _______________ = _______ items.

d Jack has 2 hammers, 6 spanners, and 9 screwdrivers.

In total, Jack has _______________ = _______ tools.

Puzzle

Each line of 3 numbers must sum to the same total.

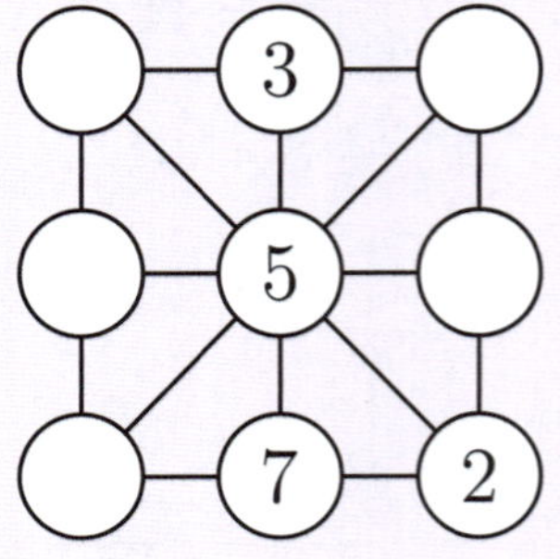

Exercise 13

Complete these additions.

a
$2 + 5 =$ ______
$12 + 5 =$ ______
$22 + 5 =$ ______
$32 + 5 =$ ______
$42 + 5 =$ ______

b
$6 + 3 =$ ______
$16 + 3 =$ ______
$26 + 3 =$ ______
$36 + 3 =$ ______
$46 + 3 =$ ______

c
$1 + 7 =$ ______
$11 + 7 =$ ______
$21 + 7 =$ ______
$31 + 7 =$ ______
$41 + 7 =$ ______

Discussion

What patterns can you see in the additions above?

Can you *explain* the patterns using blocks?

How can you *mentally* add $24 + 5$?

Exercise 14

Try to complete each addition mentally:

a $16 + 2 =$ ______
b $11 + 5 =$ ______
c $15 + 3 =$ ______

d $12 + 7 =$ ______
e $14 + 4 =$ ______
f $13 + 6 =$ ______

g $21 + 3 =$ ______
h $23 + 4 =$ ______
i $24 + 5 =$ ______

j $35 + 2 =$ ______
k $32 + 6 =$ ______
l $53 + 3 =$ ______

Exercise 15

Complete these additions.

a $3 + 8 + 2 =$ ______
b $6 + 5 + 7 =$ ______
c $5 + 9 + 3 =$ ______

d $9 + 2 + 4 =$ ______
e $8 + 4 + 5 =$ ______
f $7 + 8 + 4 =$ ______

Exercise 16

Complete these additions.

a $1 + 3 + 2 =$ _________

$10 + 30 + 20 =$ _________

b $4 + 2 + 3 =$ _________

$40 + 20 + 30 =$ _________

c $3 + 8 + 2 =$ _________

$30 + 80 + 20 =$ _________

d $6 + 7 + 1 =$ _________

$60 + 70 + 10 =$ _________

Exercise 17

Complete these additions.

a $50 + 10 + 30 =$ _________

b $60 + 20 + 30 =$ _________

c $40 + 80 + 10 =$ _________

d $90 + 50 + 30 =$ _________

e $70 + 40 + 30 =$ _________

f $80 + 60 + 50 =$ _________

Challenge

Using numbers 1 to 8, make each side total the middle number.

1

1		3
	12	
	4	2

2

3	13	7

3

	15	

Exercise 18

Complete these additions.

a

$3 + 7 =$ _________

$13 + 7 =$ _________

$23 + 7 =$ _________

$33 + 7 =$ _________

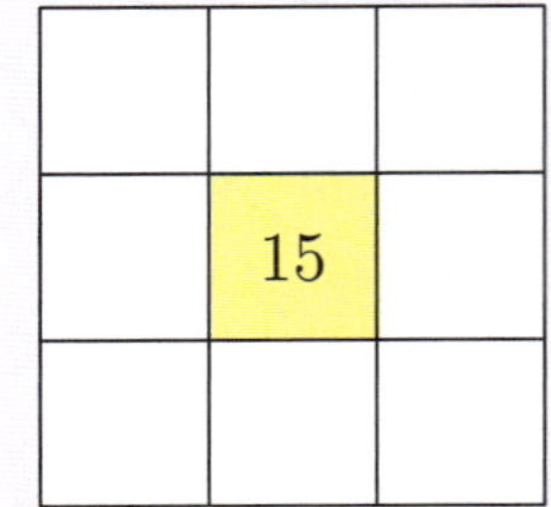

b

$8 + 2 =$ _________

$18 + 2 =$ _________

$28 + 2 =$ _________

$38 + 2 =$ _________

Discussion

What patterns can you see in **Exercise 18**?

Can you *explain* the patterns using blocks?

How can you *mentally* add $36 + 4$?

Exercise 19

Try to complete each addition mentally:

a $12 + 8 =$ _______

b $14 + 6 =$ _______

c $15 + 5 =$ _______

d $17 + 3 =$ _______

e $29 + 1 =$ _______

f $26 + 4 =$ _______

g $35 + 5 =$ _______

h $32 + 8 =$ _______

i $43 + 7 =$ _______

j $34 + 6 =$ _______

k $48 + 2 =$ _______

l $56 + 4 =$ _______

Exercise 20

Complete these additions.

a $30 + 10 + 7 =$ _______

$30 + 7 + 10 =$ _______

$7 + 30 + 10 =$ _______

b $40 + 80 + 5 =$ _______

$40 + 5 + 80 =$ _______

$5 + 40 + 80 =$ _______

Discussion

What patterns can you see in the additions above?

In which order did you add the numbers to make it easiest?

Exercise 21

Complete these additions.

a $10 + 30 + 5 =$ _______

b $7 + 40 + 10 =$ _______

c $50 + 7 + 30 =$ _______

d $30 + 60 + 5 =$ _______

e $80 + 9 + 20 =$ _______

f $6 + 70 + 40 =$ _______

Exercise 22

Complete these additions.

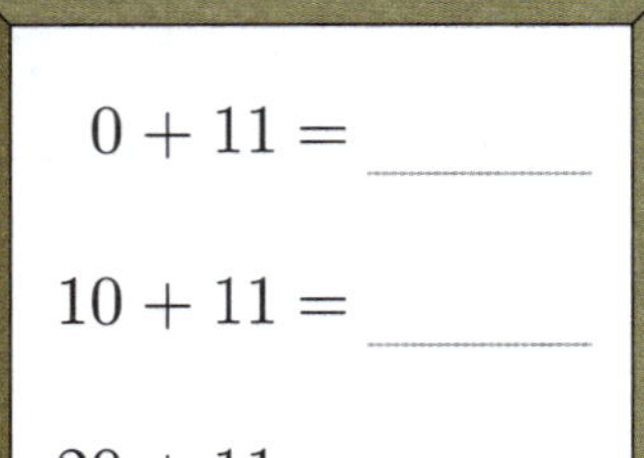

a

$0 + 11 =$ _____

$10 + 11 =$ _____

$20 + 11 =$ _____

$30 + 11 =$ _____

b

$0 + 15 =$ _____

$10 + 15 =$ _____

$20 + 15 =$ _____

$30 + 15 =$ _____

c

$0 + 17 =$ _____

$10 + 17 =$ _____

$20 + 17 =$ _____

$30 + 17 =$ _____

Exercise 23

Complete these additions.

a

$7 + 5 =$ _____

$17 + 5 =$ _____

$27 + 5 =$ _____

$37 + 5 =$ _____

b

$4 + 9 =$ _____

$14 + 9 =$ _____

$24 + 9 =$ _____

$34 + 9 =$ _____

c

$6 + 8 =$ _____

$16 + 8 =$ _____

$26 + 8 =$ _____

$36 + 8 =$ _____

Discussion

What patterns can you see in the additions above?

Can you *explain* the patterns using blocks?

How can you *mentally* add $25 + 7$?

Exercise 24

Complete these additions.

a $16 + 8 =$ _____

b $16 + 9 =$ _____

c $15 + 7 =$ _____

d $15 + 9 =$ _____

e $18 + 7 =$ _____

f $19 + 8 =$ _____

Exercise 25

Complete these additions.

a $19 + 9 = $ _______

b $14 + 9 = $ _______

c $18 + 8 = $ _______

d $17 + 9 = $ _______

e $19 + 7 = $ _______

f $18 + 9 = $ _______

g $23 + 8 = $ _______

h $24 + 7 = $ _______

i $25 + 7 = $ _______

j $25 + 8 = $ _______

k $26 + 7 = $ _______

l $26 + 9 = $ _______

m $35 + 9 = $ _______

n $36 + 8 = $ _______

o $37 + 7 = $ _______

Exercise 26

Complete these additions.

a $7 + 6 + 9 = $ _______

b $5 + 8 + 9 = $ _______

c $9 + 4 + 7 = $ _______

d $8 + 7 + 6 = $ _______

e $6 + 9 + 9 = $ _______

f $7 + 8 + 7 = $ _______

Exercise 27

Write an addition to solve each problem:

a Celine has 26 dolls. She collects 3 more.

In total, Celine has _______ + _______ = _______ dolls.

b Hung has 6 marbles, 9 toy cars, and 3 blocks.

Altogether, Hung has _______ + _______ + _______ = _______ toys.

c David has 17 cows in one field and 8 cows in another.

In total, David has _______ + _______ = _______ cows.

d Katalin has 22 small containers and 9 large containers.

Altogether, Katalin has _______ + _______ = _______ containers.

e Mr Pierson has 8 basketballs, 6 footballs, and 9 baseballs.

In total, Mr Pierson has _______ + _______ + _______ = _______ balls.

Game

Click the icon to practise your addition.

Exercise 28

Complete these additions.

a $5 + 13 =$ _____   **b** $7 + 21 =$ _____   **c** $4 + 16 =$ _____

d $3 + 27 =$ _____   **e** $6 + 17 =$ _____   **f** $9 + 23 =$ _____

g $4 + 32 =$ _____   **h** $8 + 42 =$ _____   **i** $7 + 35 =$ _____

j $6 + 29 =$ _____   **k** $5 + 55 =$ _____   **l** $3 + 48 =$ _____

$$2 \quad + \quad 3 \quad \neq \quad 6$$

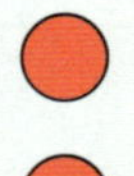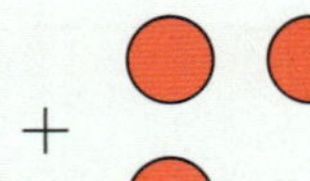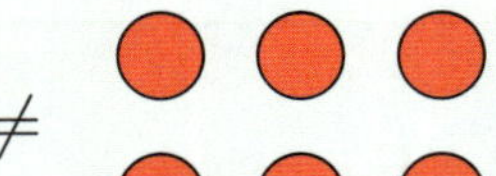

Exercise 29

Complete using $=$ or $\neq$:

a $7 + 2$ _____ 9 **b** $3 + 8$ _____ 12 **c** $8 + 5$ _____ 13

d $20 + 5$ _____ 25 **e** $30 + 4$ _____ 43 **f** $70 + 40$ _____ 74

g $5 + 40$ _____ 54 **h** $9 + 80$ _____ 89 **i** $30 + 50$ _____ 80

j $3 + 7 + 2$ _____ 11 **k** $12 + 6$ _____ 18 **l** $23 + 5$ _____ 27

m $30 + 4 + 20$ _____ 54 **n** $6 + 8 + 5$ _____ 19 **o** $37 + 3$ _____ 30

p $26 + 9$ _____ 35 **q** $38 + 5$ _____ 33 **r** $6 + 9 + 5$ _____ 20

s $70 + 80 + 30$ _____ 180 **t** $15 + 8$ _____ 21 **u** $8 + 7 + 8$ _____ 23

Exercise 30

Describe what is happening in each sequence:

a { 4 } → { 8 } → { 12 } → { 16 } → { 20 }

I start with _______ , and add _______ each time.

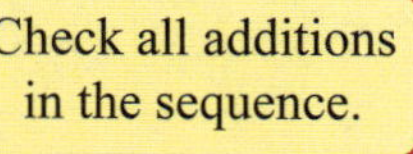

b △ 1 → △ 6 → △ 11 → △ 16 → △ 21

____

c 3 → 7 → 11 → 15 → 19

____

d 12 → 18 → 24 → 30 → 36

____

Exercise 31

Add 2 each time:

a 2 → ☐ → ☐ → ☐ → ☐ → ☐ → ☐

b 7 → ☐ → ☐ → ☐ → ☐ → ☐ → ☐

Add 3 each time:

c 4 → ☐ → ☐ → ☐ → ☐ → ☐ → ☐

d 9 → ☐ → ☐ → ☐ → ☐ → ☐ → ☐

Add 5 each time:

e 1 → ☐ → ☐ → ☐ → ☐ → ☐ → ☐

Add 7 each time:

f 3 → ☐ → ☐ → ☐ → ☐ → ☐ → ☐

Puzzle

Use the numbers 1 to 12 once each to complete these sums:

$$\boxed{} + \boxed{} + \boxed{} = 18 \qquad \boxed{} + \boxed{} + \boxed{} = 20$$

$$\boxed{} + \boxed{} + \boxed{} = 19 \qquad \boxed{} + \boxed{} + \boxed{} = 21$$

Puzzle

Maizie has 8 cards that are numbered 1 to 8.

She has chosen 4 cards that total 20. What cards could they be?

There are seven solutions. Can you find them all?

Write your answers below.

To add numbers with more than one digit, we can use **column addition**.
We write the numbers in columns so the place values line up. We then add each column, starting with the units.

	T	U
	2	1
+	3	8
	5	9

1 Set out your addition.

2 Add the units:

$$1 + 8 = 9$$

3 Add the tens:

$$2 + 3 = 5$$

So, $21 + 38 = 59$

Exercise 32

Complete these additions.

a

	T	U
	1	5
+	1	2

b

	T	U
	2	1
+	1	5

c

	T	U
	2	2
+	1	6

d

	T	U
	2	7
+	2	1

e

	T	U
	3	2
+	2	5

f

	T	U
	3	4
+	2	3

g

	T	U
	3	0
+	3	9

h

	T	U
	4	3
+	2	5

Exercise 33

Set out these additions and find the answer.

a $16 + 20$

b $24 + 31$

c $32 + 16$

d $45 + 30$

e $54 + 24$

f $37 + 51$

g $45 + 12$

h $71 + 24$

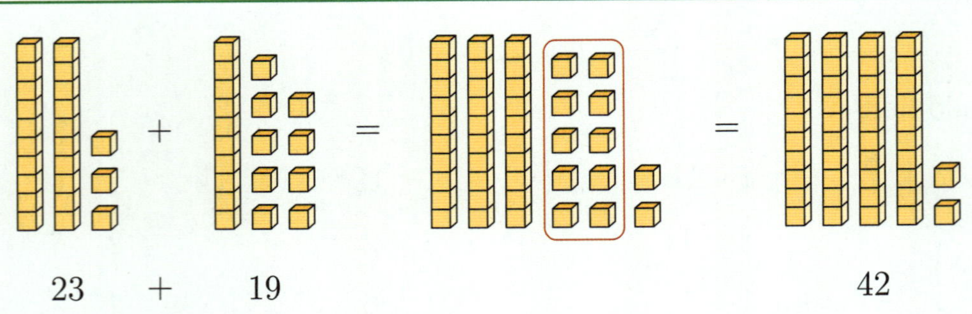

$$23 \quad + \quad 19 \qquad\qquad\qquad 42$$

If there are 10 or more units, we **carry** a ten into the tens column.

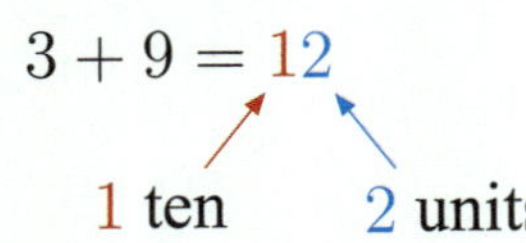

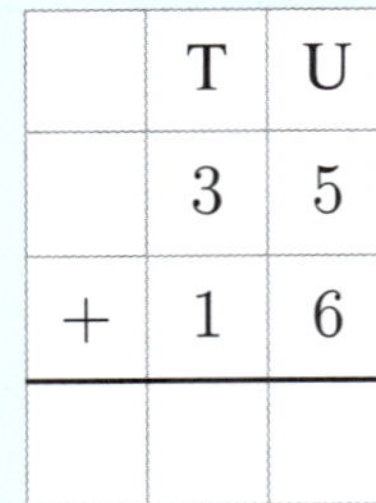

1 Set out your addition.

2 Add the units:

$$3 + 9 = 12$$

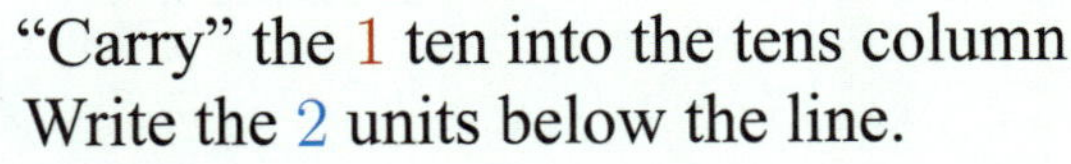

1 ten   2 units

"Carry" the 1 ten into the tens column.
Write the 2 units below the line.

3 Add the tens:

$$2 + 1 + 1 = 4$$

So, $23 + 19 = 42$.

Exercise 34

Complete these additions.

a

	T	U
	2	5
+	1	6

b

	T	U
	3	4
+	1	9

c

	T	U
	2	7
+	1	4

d

	T	U
	3	5
+	1	6

e

	T	U
	4	6
+	2	6

f

	T	U
	3	9
+	2	1

g

	T	U
	4	8
+	2	3

h

	T	U
	3	6
+	3	7

i

	T	U
	4	5
+	3	9

j

	T	U
	2	8
+	2	7

k

	T	U
	5	5
+	3	8

l

	T	U
	2	4
+	6	7

Exercise 35

Set out these additions and find the answer.

a 26 + 19

	T	U

b 35 + 27

	T	U

c 36 + 35

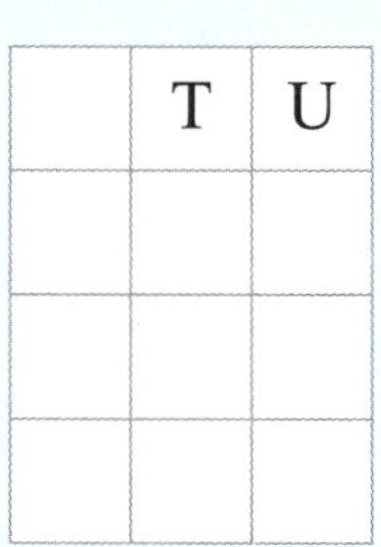

d 41 + 39

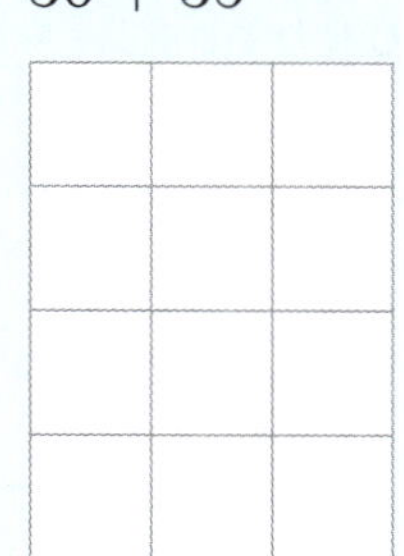

e 25 + 66

f 54 + 28

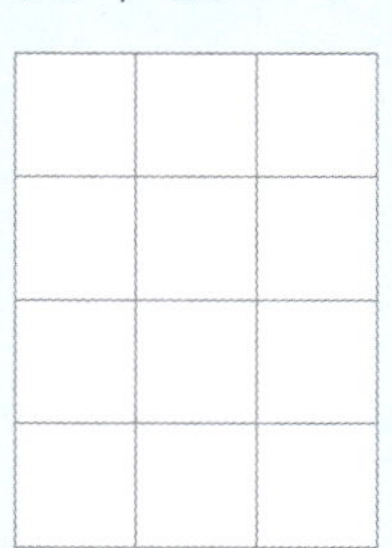

g 57 + 35

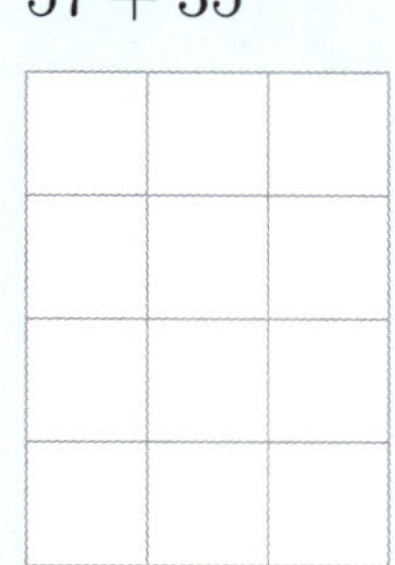

h 34 + 59

i 68 + 24

j 54 + 29

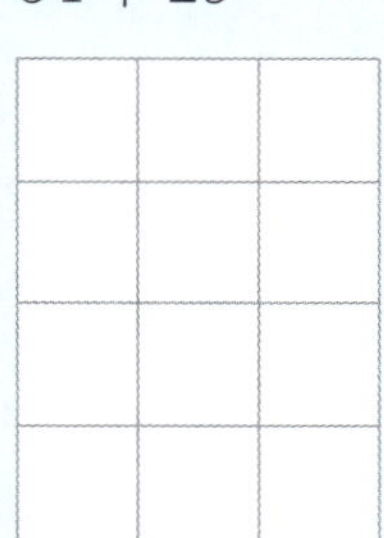

k 32 + 49

l 74 + 18

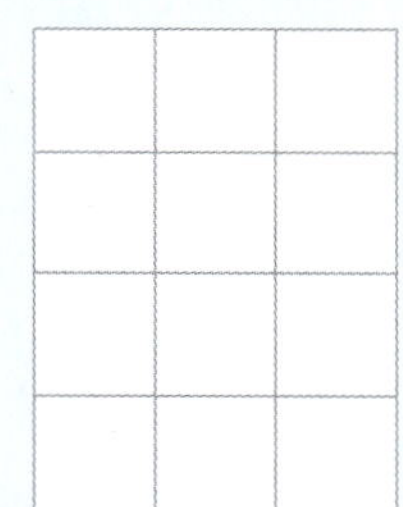

Exercise 36

Circle the words that tell you to **add**, then work out the answer.

a Sam counted 24 butterflies on Monday, and 15 on Tuesday.

How many butterflies did Sam count altogether?

Sam counted ☐ butterflies.

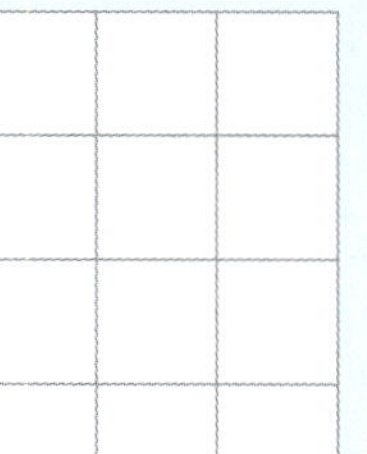

b Clare has 35 apples trees. Jonathon has 28 apple trees.

Find the total number of apple trees.

There are ☐ apple trees.

Puzzle

What do you call a cow on a trampoline?

Answer each addition. Each answer will give you a letter and help you to find the answer.
Write the matching letter in each box.

a

	T	U
	2	9
+	1	7

b

	T	U
	5	7
+	2	8

c $43 + 37$

d $38 + 27$

e

	T	U
	4	5
+	4	7

f

	T	U
	3	4
+	5	9

g $19 + 48$

h $22 + 24$

i

	T	U
	2	8
+	6	4

j

	T	U
	2	4
+	2	7

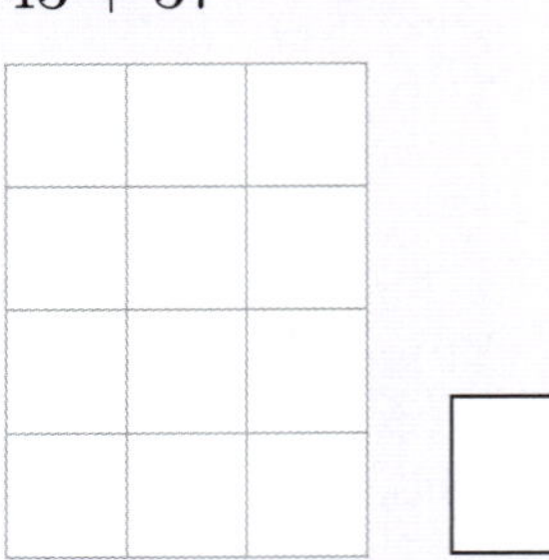

H	L	B	T	M	P	S
67	65	54	75	85	48	93
A	O	N	K	J	E	I
46	89	72	92	63	51	80

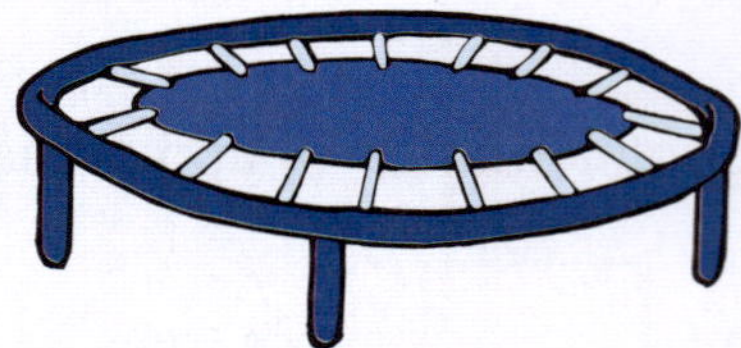

Answer: ________ ________ ________ ________ ________   ________ ________ ________ ________ ________ !

Exercise 37

Solve these problems.

a Jane scored 38 points in Saturday's game, and 26 points on Sunday.

How many points did Jane score in total?

In total, Jane scored ☐ points.

b Lachlan has 49 football cards, and William has 37 football cards.

Find the total number of football cards.

There are ☐ football cards.

c Annie and Tia are twins. They love to swim.

In one year, Annie won 16 medals and 18 certificates.

Tia won 17 medals and 19 certificates.

Altogether, the twins won

☐ medals and ☐ certificates.

In total, the twins won ☐ prizes.

If there are 10 or more tens, we **carry** a hundred into the hundreds column.

$57 + 82$

	H	T	U
		5	7
+	₁	8	2
	1	3	9

1 Set out your addition.

2 Add the units:

$7 + 2 = 9$

3 Add the tens:

$5 + 8 = 13$

"Carry" the 1 into the hundreds column.
Write the 3 below the line.

4 Write the 1 hundred below the line.

So, $57 + 82 = 139$.

Exercise 38

Complete these additions.

a
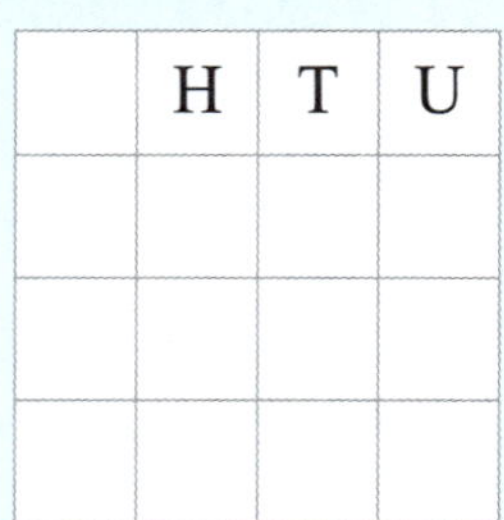

	H	T	U
		6	4
+		7	3

b

	H	T	U
		9	1
+		8	5

c

	H	T	U
		4	8
+		7	6

Exercise 39

Set out these additions and find the answer.

a $37 + 82$

H	T	U

b $61 + 57$

H	T	U

c $83 + 19$

d $94 + 38$

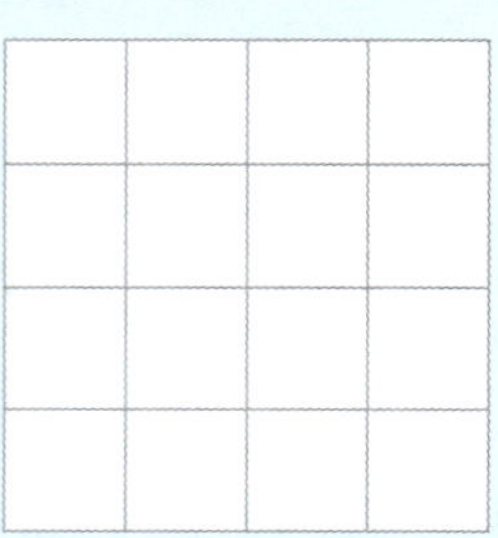

e $66 + 54$

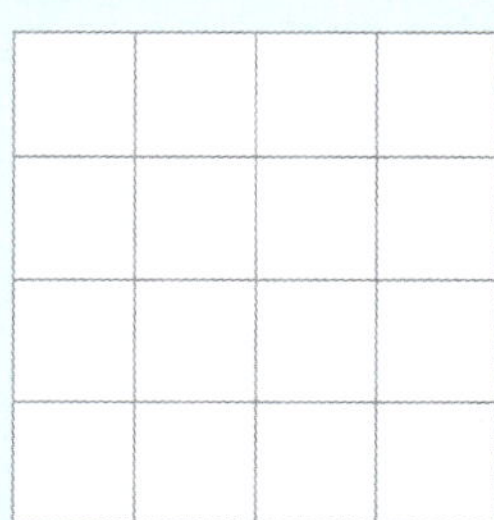

f $78 + 85$

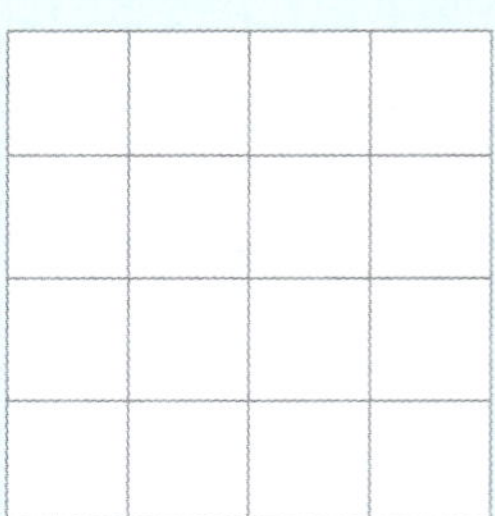

Revision

1 a Add together 4 and 9. ________   **b** Find the sum of 8 and 5. ________

   c Find the total of 7 and 4. ________   **d** What is 6 plus 6? ________

2 Complete these additions.

   a $30 + 40 =$ ________   **b** $50 + 20 =$ ________   **c** $50 + 70 =$ ________

   d $40 + 90 =$ ________   **e** $70 + 70 =$ ________   **f** $80 + 30 =$ ________

3 Complete these additions.

   a $6 +$ ________ $= 10$ **b** ________ $+ 5 = 13$ **c** $7 +$ ________ $= 15$

   d ________ $+ 9 = 69$ **e** $80 +$ ________ $= 90$ **f** ________ $+ 50 = 120$

4 Complete using $=$ or $\neq$:

a $9 + 8$ _____ 17 **b** $50 + 6$ _____ 65 **c** $8 + 5$ _____ $4 + 7$

d $3 + 70$ _____ $70 + 3$ **e** $20 + 60$ _____ 80 **f** $13 + 9$ _____ $2 + 20$

5 Add mentally:

a $14 + 3 =$ _____   **b** $13 + 6 =$ _____   **c** $15 + 4 =$ _____

d $31 + 8 =$ _____   **e** $26 + 2 =$ _____   **f** $33 + 4 =$ _____

6 Complete each addition:

a $4 + 2 + 5 =$ _____   **b** $5 + 5 + 6 =$ _____   **c** $2 + 7 + 4 =$ _____

d $6 + 7 + 4 =$ _____   **e** $7 + 5 + 3 =$ _____   **f** $4 + 8 + 6 =$ _____

7 Complete each addition:

a $30 + 40 + 20 =$ _____   **b** $60 + 50 + 30 =$ _____

c $50 + 6 + 20 =$ _____   **d** $9 + 30 + 80 =$ _____

8 Add mentally:

a $20 + 13 =$ _____   **b** $40 + 13 =$ _____   **c** $70 + 13 =$ _____

d $19 + 6 =$ _____   **e** $18 + 8 =$ _____   **f** $17 + 5 =$ _____

g $16 + 9 =$ _____   **h** $14 + 7 =$ _____   **i** $15 + 8 =$ _____

9 Complete these additions.

a $28 + 4 =$ _____   **b** $23 + 9 =$ _____   **c** $45 + 6 =$ _____

d $37 + 8 =$ _____   **e** $32 + 9 =$ _____   **f** $58 + 7 =$ _____

10 Complete using $=$ or $\neq$:

a $19 + 8$ _____ 27 **b** $23 + 9$ _____ 31 **c** $18 + 9$ _____ 29

d $25 + 8$ _____ 32 **e** $24 + 7$ _____ 31 **f** $23 + 9$ _____ 32

11 Add 6 each time: $3 \rightarrow \boxed{} \rightarrow \boxed{} \rightarrow \boxed{} \rightarrow \boxed{} \rightarrow \boxed{}$

12 Describe what is happening in the sequence.

$5 \rightarrow 12 \rightarrow 19 \rightarrow 26 \rightarrow 33$

I start with _____ , and add _____ each time.

13 Sebastian has 6 fish, 5 birds, and 3 cats.

In total, Sebastian has _______ + _______ + _______ = _______ pets.

14 Complete these additions.

a
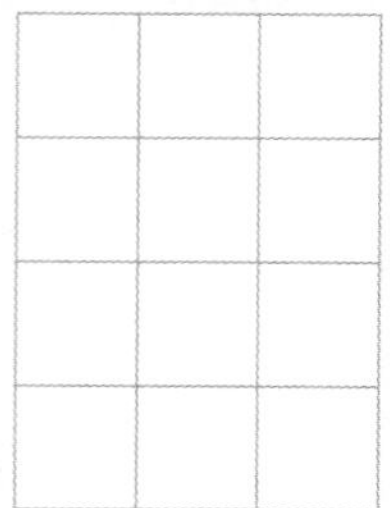

	T	U
	4	3
+	2	6

b

	T	U
	3	5
+	3	2

c

	T	U
	4	9
+	3	6

d

	T	U
	5	7
+	3	9

15 Set out these additions and find the answer.

a $62 + 15$

b $34 + 53$

c $32 + 29$

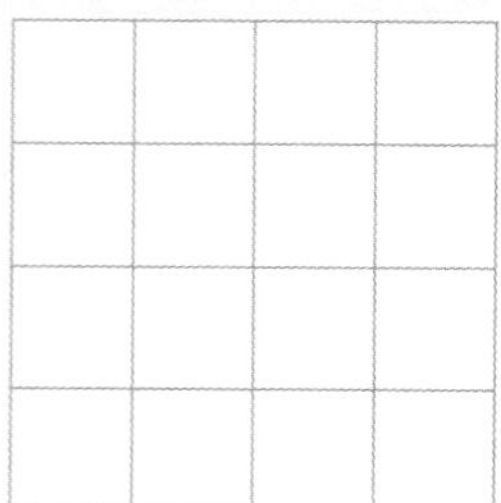

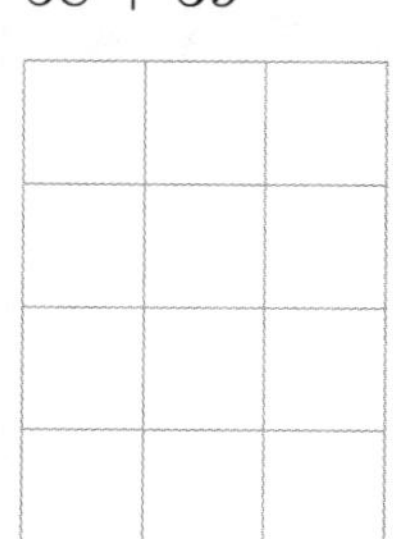

d $65 + 46$

e $58 + 39$

f $68 + 75$

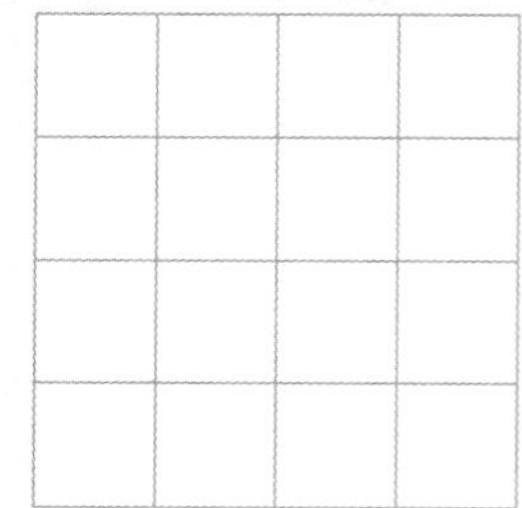

16 Caroline picked 24 oranges and 19 lemons.

How many fruit did she pick in total?

Caroline picked [ ] fruit in total.

17 a My team scored 38 points last week and 47 points this week.

In total, my team scored [ ] points.

b If my team scores 43 points in its next match,

its new total will be [ ] points.

A **straight line** never changes direction.   A **curve** is not straight.

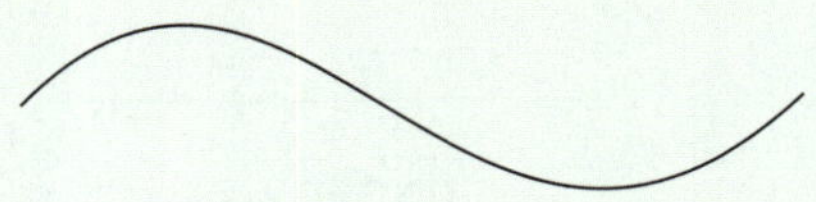

A shape is **2-dimensional** if it can be drawn on a **surface**.

Exercise 1

These 2-dimensional shapes have straight sides.

a Practise your spelling. Copy the name of each shape.

b Complete the number of sides.

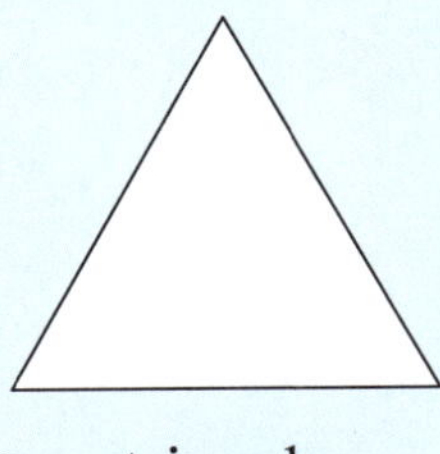

triangle

____

_______ sides

quadrilateral

____

_______ sides

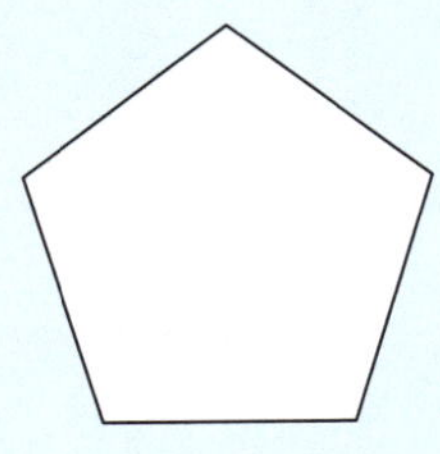

pentagon

____

_______ sides

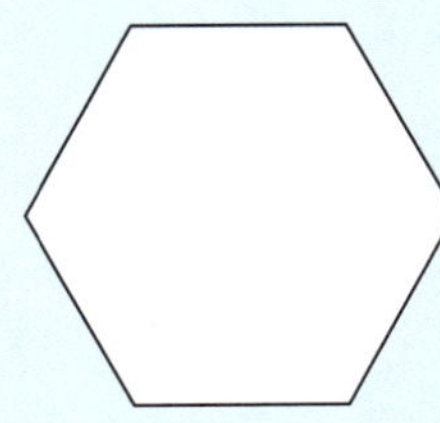

hexagon

____

_______ sides

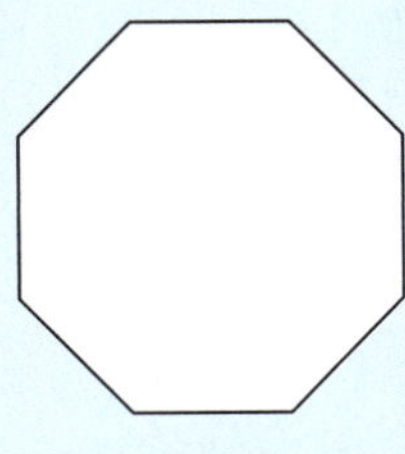

octagon

____

_______ sides

Exercise 2

Here are some special quadrilaterals.

rectangle

square

_______________________   _______________________

a Copy the name of each shape.

b Complete:

A rectangle is special because:

- opposite sides have the same _______________

- all of its corners look the _____________ .

A square is special because:

- __________ of its sides have the same _______________

- all of its corners look the _____________ .

Exercise 3

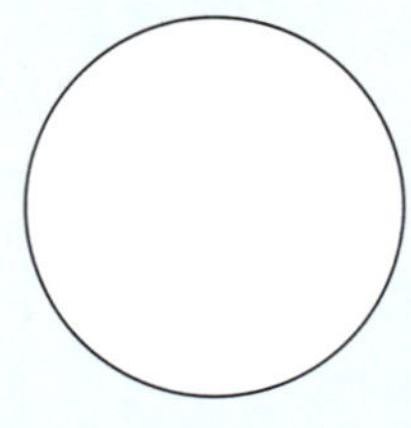

circle

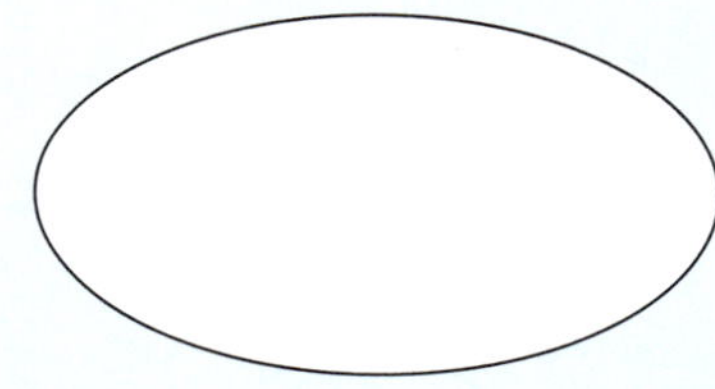

ellipse

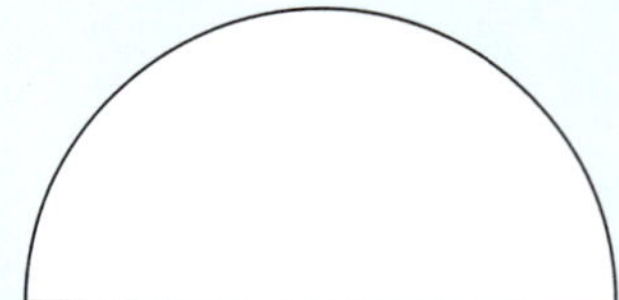

semi-circle

_______________   _______________   _______________

a Copy the name of each shape.

b Complete:

A _______________ and an _______________ are drawn with a single curve.

A semi-circle has a curve and a _______________ line.

© HAESE MATHEMATICS 2024

Activity

Sort these shapes into groups.

a

b

c

d

e

f

g

h

i

j

k

l

m

n

o

p

q

r

s

t

u

triangles	quadrilaterals	pentagons
hexagons	octagons	shapes with curves

Exercise 4

List the shapes you can see.

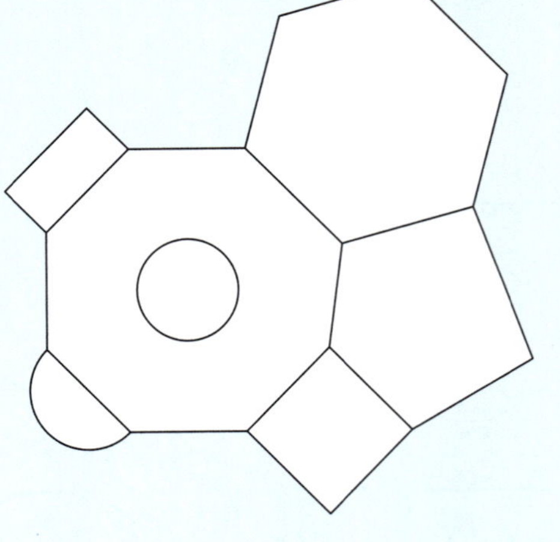

____

____

____

____

____

____

Exercise 5

Draw each shape. Use your ruler if needed.

a a square

b a semi-circle

c a hexagon

d an ellipse

Listening Activity

Click and listen carefully to the instructions.

1 ___________________

2 ___________________

3 ___________________

4 ___________________

5

6

Exercise 6

How would you *describe* a rectangle to someone?

____

____

Puzzle

Without lifting your pencil, draw *six* lines which create *two* rectangles.

Puzzle

How many triangles can you find?

____

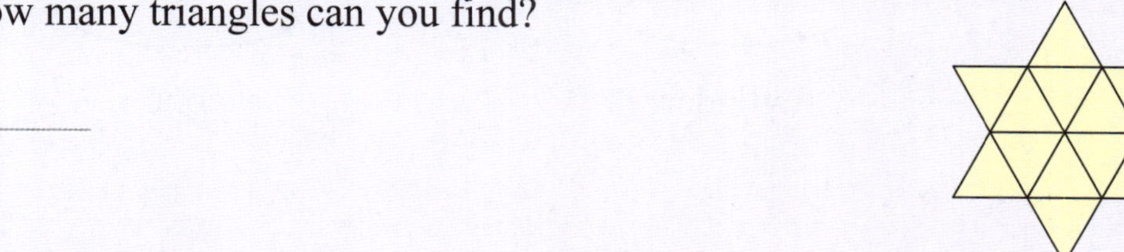

A **solid** is a 3-dimensional shape which takes up **space**.

We can *draw* solids as 2-dimensional shapes on a page.

Any flat surface of a solid is called a **face**.

Any line where two faces meet is called an **edge**.

Any point where three faces meet is called a **corner**.

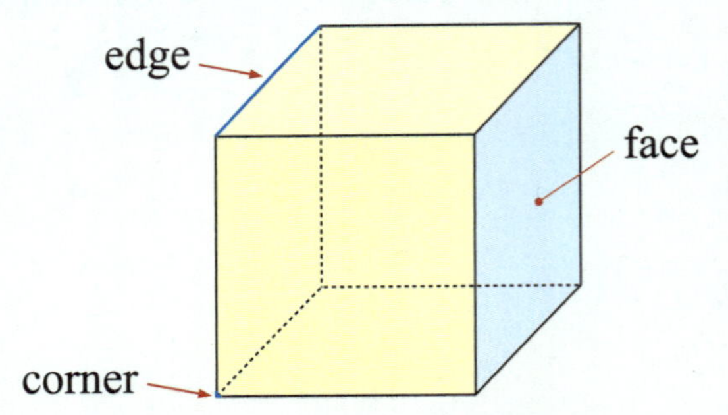

Exercise 7

Here are some 3-dimensional solids. Practise your spelling.

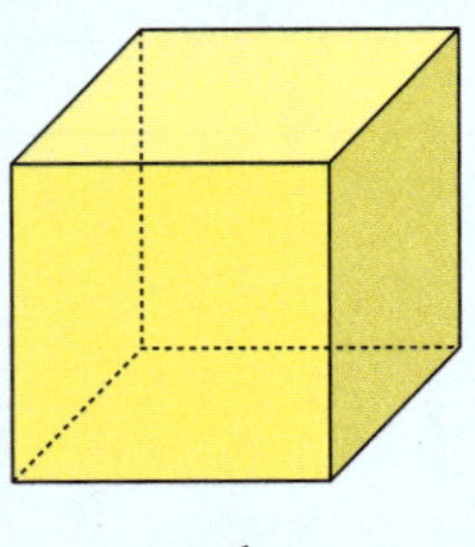

cube

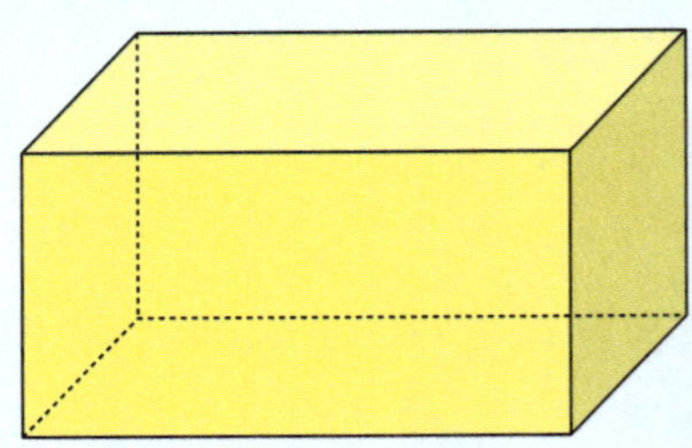

rectangular prism

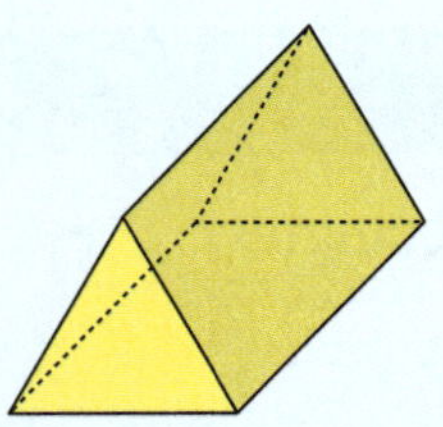

triangular prism

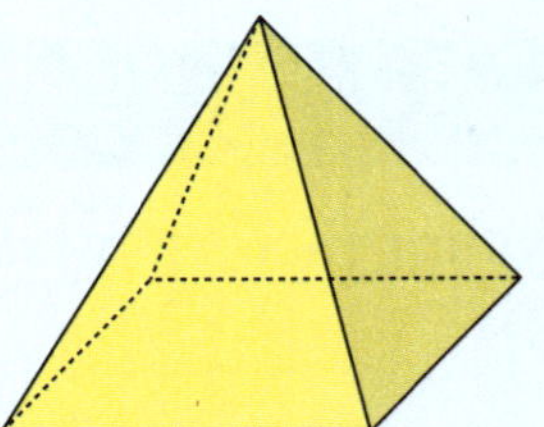

square-based pyramid

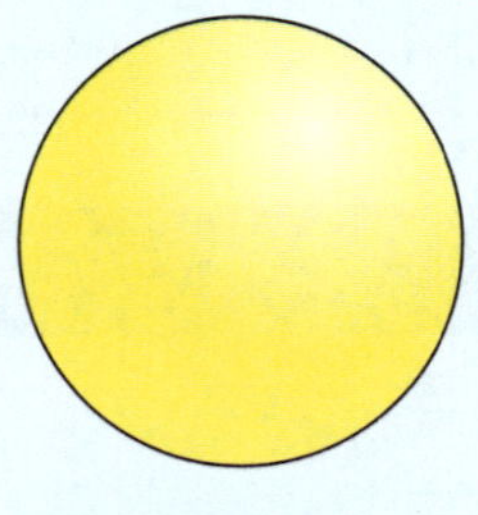

sphere

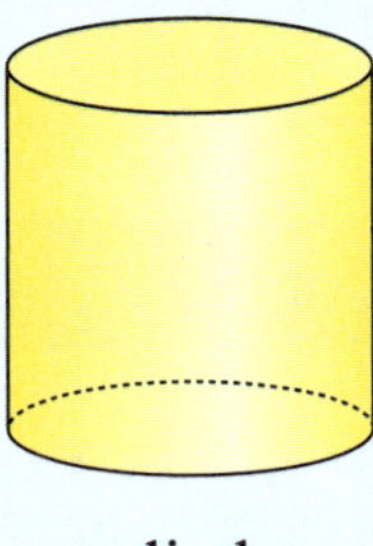

cylinder

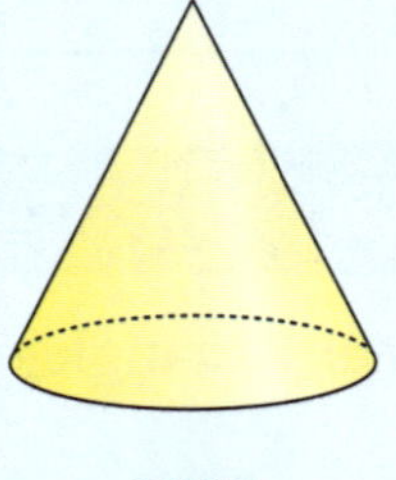

cone

Click on each solid to view it from other angles.

Exercise 8

Complete the table.

Solid	*Faces*	*Edges*	*Corners*
cube			
rectangular prism			
triangular prism			
square-based pyramid			

Exercise 9

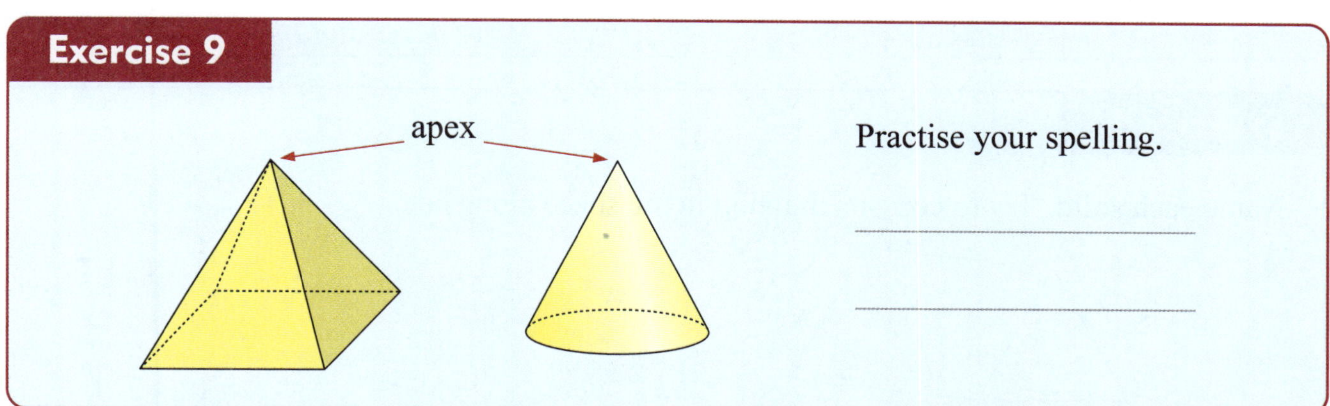

Practise your spelling.

____

____

Discussion

This is a cylinder. How can you *describe* a cylinder?

____

____

Where are cylinders used?

____

____

____

Exercise 10

Complete each sentence:

a Three solids which have curved surfaces are a _________________ ,

a _________________ , and a _________________ .

b A triangular prism has 2 _________________ faces and

3 _________________ faces.

c A square-based pyramid has 1 _________________ face and _______ triangular

faces. The point where the _________________ faces meet is called the

_________________ .

d A cube has _______ _________________ faces.

Exercise 11

Name each solid. Try to copy its drawing in the space alongside.

a

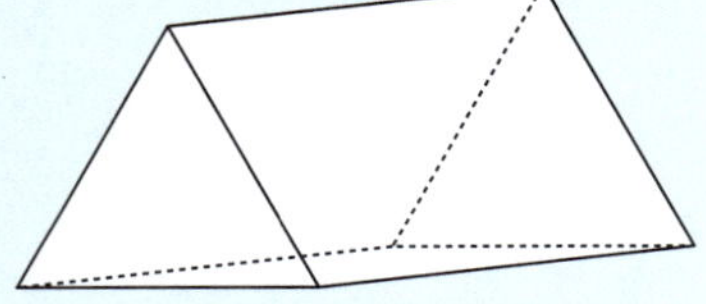

____

b

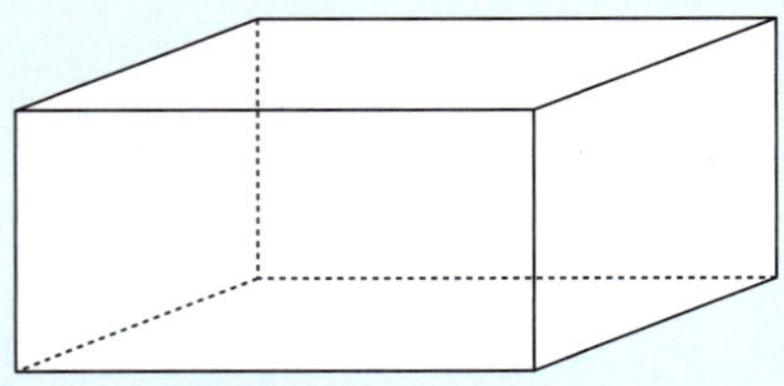

____

c

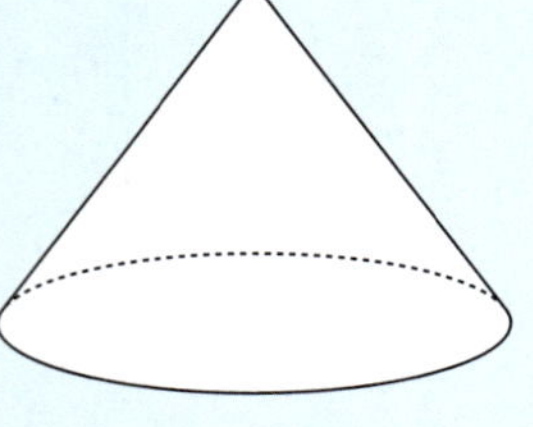

____

Activity   Name that solid

You will need: a partner, a collection of different solids

What to do:

1   Place your hands behind your back. Your partner will choose a solid and place it into your hands without showing you.

2   Guess the solid, and ask your partner to take the solid away again. Write your guess in the table below.

3   Ask your partner to reveal the solid. Fill in the second column in the table. If you did not guess correctly, feel the solid again.

4   Repeat with 4 more solids.

5   Swap with your partner to give them a turn. Use 5 solids in a different order.

My guess	*The solid was a ....*

Exercise 12

Name each solid. Try to copy its drawing in the space alongside.

a

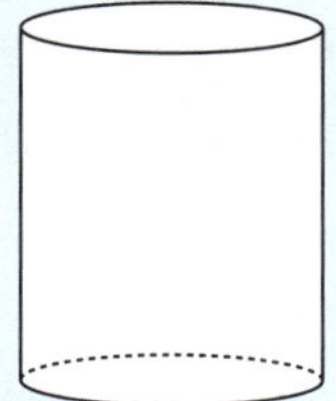

____

b

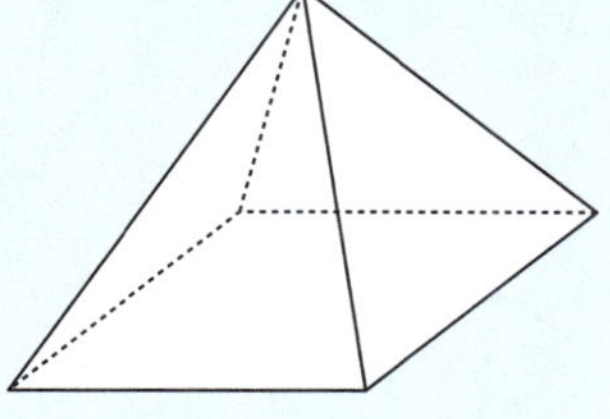

____

A **net** is a 2-dimensional shape which can be folded to make a 3-dimensional solid.

Think about the *surface* of a cube. If you unfold the surface, you will see its **net**.

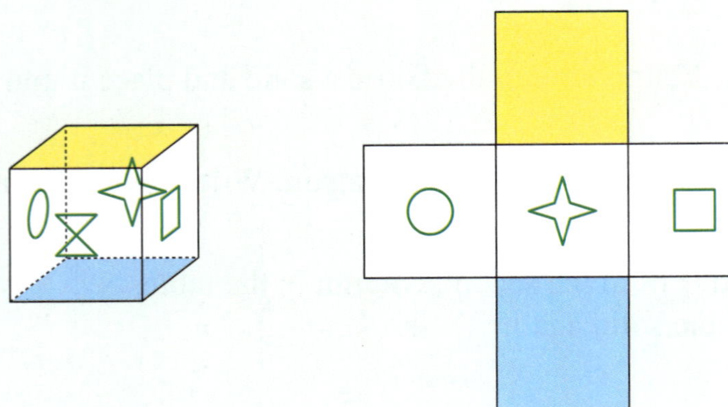

Click on the icon to see how the net is folded into the cube.

Activity   Making solids

You will need:

printed sheets, coloured pencils, scissors, tape

What to do:

1 Print the nets.

2 Colour each face of each solid a different colour.

3 Cut around the *thick* black lines.

4 Fold along each red line.

5 Use tape to join the sides.

Puzzle

For which net will the yellow face be opposite the blue face?

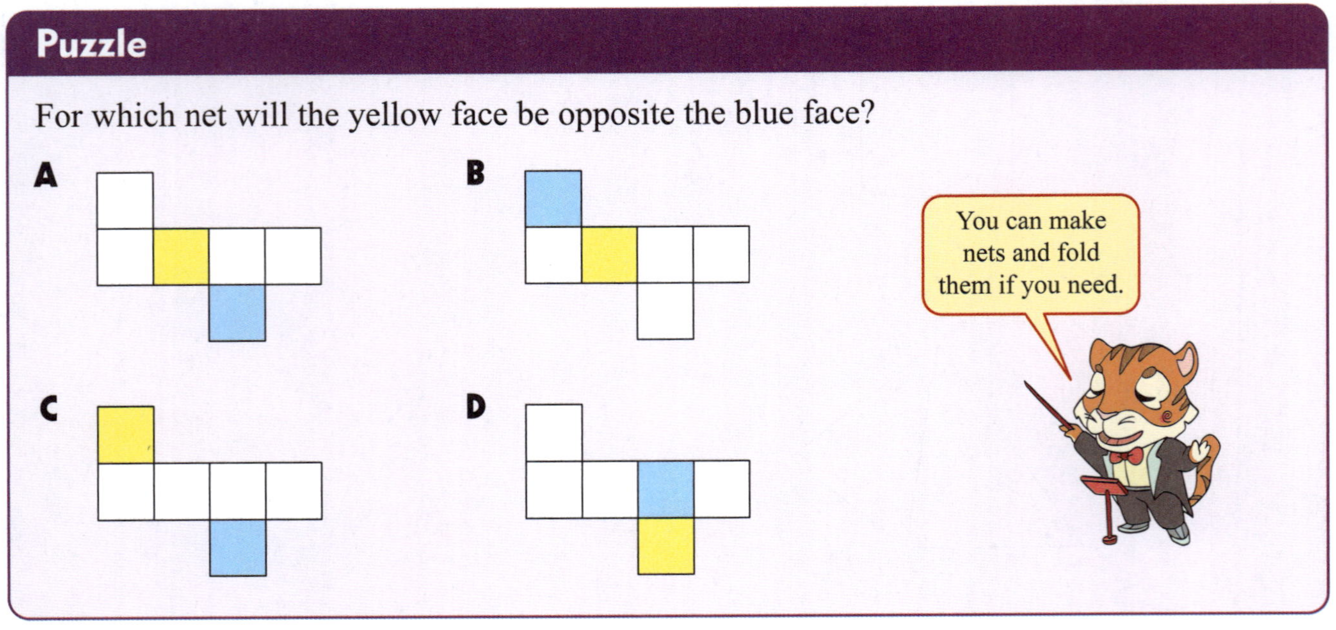

Exercise 13

Match each net with its solid:

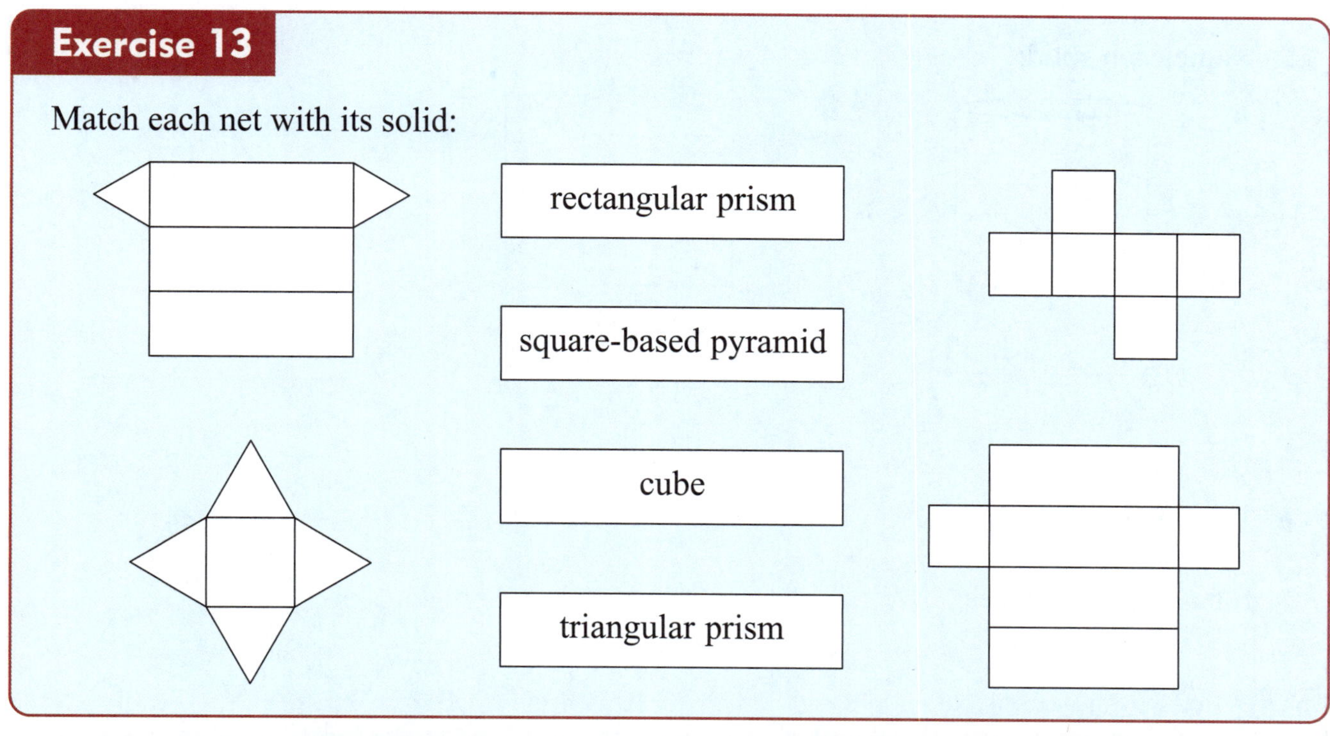

rectangular prism

square-based pyramid

cube

triangular prism

Revision

1 Name each shape:

a   **b**   **c**

__________________   __________________   __________________

2 How would you describe a pentagon to someone?

____

____

3 Draw a:

    a triangle   **b** circle

4 Complete each sentence:

    a A hexagon has __________ sides.

    b An __________________________ has 8 sides.

    c A circle and an ellipse are both drawn with a single __________________.

5 Name each solid:

a

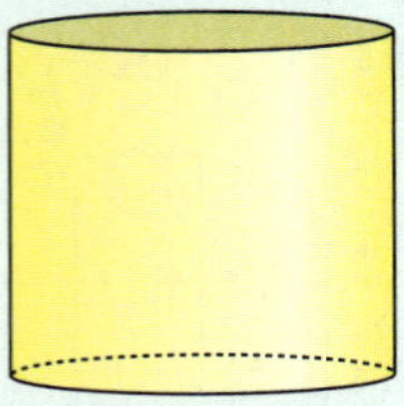

b

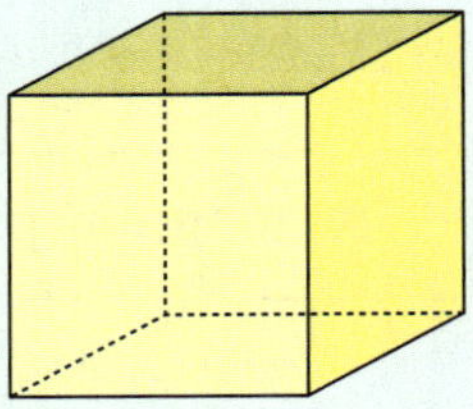

c

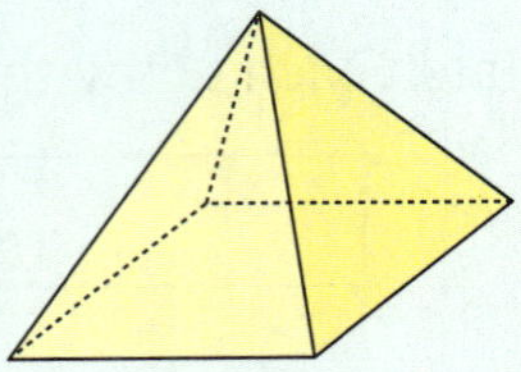

____

____

____

6 This shape is a _________________________ .

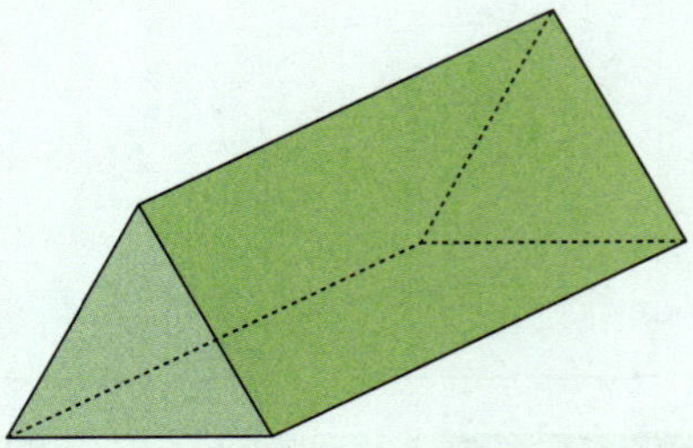

It has _______ faces,

_______ edges,

and _______ corners.

7 Complete:

  a A sphere has a single _________________________ surface.

  b A rectangular prism has six _________________________ faces.

  c The point at the top of a cone is called the _________________________ .

8 What shape can this net be folded into?

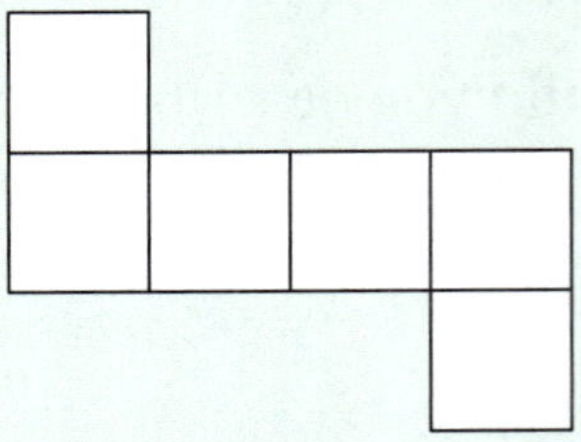

____

Discussion

$$8 - 3 = 5$$

"eight minus three equals five"

What words tell us that we need to subtract?

______________________   ______________________   ______________________

______________________   ______________________   ______________________

____

To find $13 - 5$, we start at 13 then subtract 5.

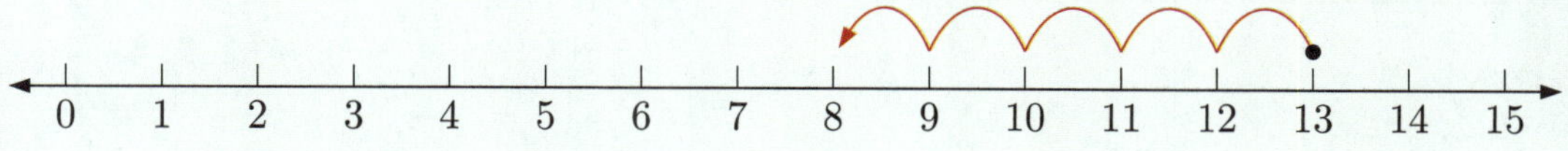 

$13 - 5 = 8$

Exercise 1

Use the number line to help you subtract:

a

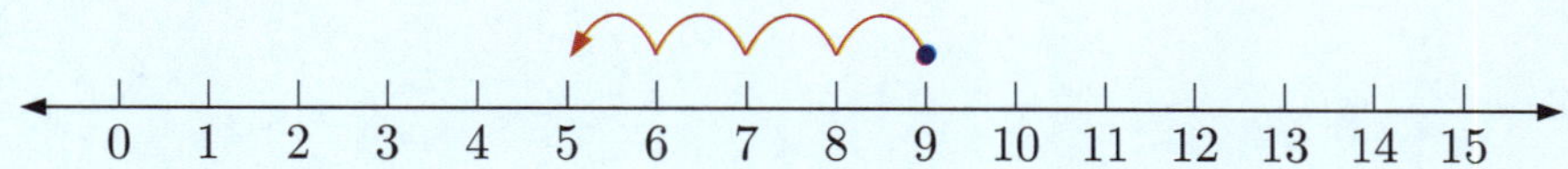

$9 - 4 = $ ______

b

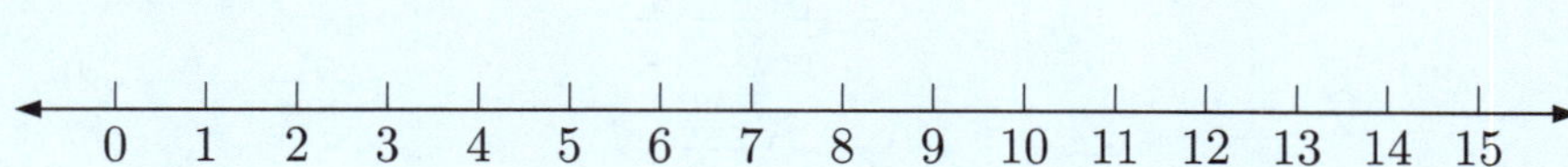

$12 - 5 = $ ______

c

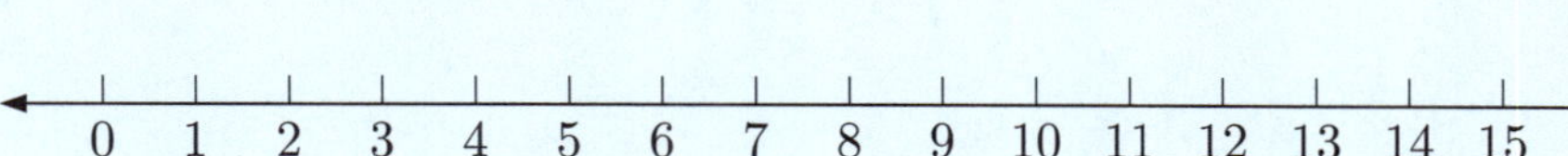

$14 - 8 = $ ______

Exercise 2

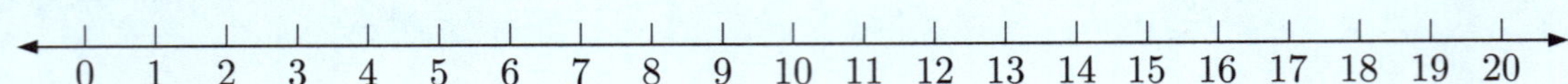

Use the number line to help you subtract:

a $8 - 3 = $ _______

b $10 - 6 = $ _______

c $15 - 2 = $ _______

d $12 - 4 = $ _______

e $13 - 10 = $ _______

f $16 - 5 = $ _______

g $18 - 11 = $ _______

h $16 - 9 = $ _______

i $20 - 8 = $ _______

Addition and subtraction are related.

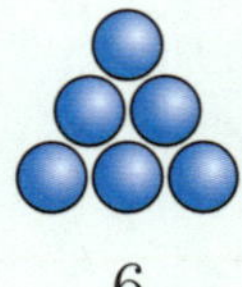
6

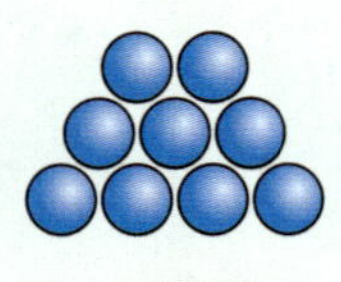
9

There are 15 balls in total.

$$6 + 9 = 15$$
$$9 + 6 = 15$$
$$15 - 6 = 9$$
$$15 - 9 = 6$$

Exercise 3

Complete:

a $8 + 5 = $ _______

$5 + 8 = $ _______

_______ $- 5 = 8$

_______ $- 8 = 5$

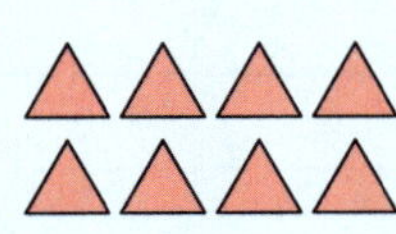

8

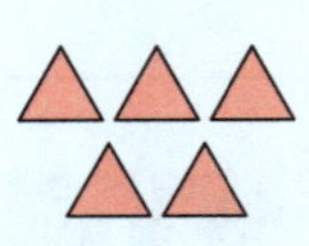
5

b   _______ $+$ _______ $=$ _______

_______ $+$ _______ $=$ _______

_______ $-$ _______ $=$ _______

_______ $-$ _______ $=$ _______

_______   _______

Exercise 4

Find the missing numbers.

a $7 = 3 +$ ▢

so $7 - 3 =$ ▢

b $6 = 4 +$ ▢

so $6 - 4 =$ ▢

c $10 = 7 +$ ▢

so $10 - 7 =$ ▢

Exercise 5

Use the addition to write *two* subtractions.

a
$$3 + 6 = 9$$
so $9 - \underline{\qquad} = \underline{\qquad}$

and $9 - \underline{\qquad} = \underline{\qquad}$

b
$$7 + 5 = 12$$
so $12 - \underline{\qquad} = \underline{\qquad}$

and $12 - \underline{\qquad} = \underline{\qquad}$

c
$$5 + 4 = \underline{\qquad}$$
so $\underline{\qquad} - 5 = \underline{\qquad}$

and $\underline{\qquad} - 4 = \underline{\qquad}$

d
$$8 + 3 = \underline{\qquad}$$
so $\underline{\qquad} - 8 = \underline{\qquad}$

and $\underline{\qquad} - 3 = \underline{\qquad}$

e
$$9 + 5 = \underline{\qquad}$$
so $\underline{\qquad} - 9 = \underline{\qquad}$

and $\underline{\qquad} - 5 = \underline{\qquad}$

f
$$7 + 8 = \underline{\qquad}$$
so $\underline{\qquad} - 7 = \underline{\qquad}$

and $\underline{\qquad} - 8 = \underline{\qquad}$

Mental subtractions are done without a pencil and paper, or apparatus.

Exercise 6

Write down the result of each subtraction:

a $3 - 2 = \underline{\qquad}$

b $6 - 3 = \underline{\qquad}$

c $7 - 1 = \underline{\qquad}$

d $8 - 5 = \underline{\qquad}$

e $7 - 5 = \underline{\qquad}$

f $9 - 6 = \underline{\qquad}$

g $9 - 2 = \underline{\qquad}$

h $8 - 3 = \underline{\qquad}$

i $6 - 6 = \underline{\qquad}$

Exercise 7

Write down the result of each subtraction:

a $10 - 3 =$ _____

b $13 - 3 =$ _____

c $10 - 5 =$ _____

d $11 - 4 =$ _____

e $14 - 4 =$ _____

f $13 - 6 =$ _____

g $15 - 6 =$ _____

h $18 - 8 =$ _____

i $12 - 7 =$ _____

j $17 - 9 =$ _____

k $19 - 9 =$ _____

l $16 - 8 =$ _____

Exercise 8

Complete:

a 10 minus $8 =$ _____

b 8 take away $4 =$ _____

c 11 subtract $7 =$ _____

d Find the difference: $9 - 3 =$ _____

e 14 subtract $8 =$ _____

f Find the difference: $12 - 9 =$ _____

g Subtract 6 from 8: $8 - 6 =$ _____

h Take 4 away from 9:   _____ $-$ _____ $=$ _____

i 7 less than 15:   _____ $-$ _____ $=$ _____

The **difference** between two numbers is the distance between them on the number line.

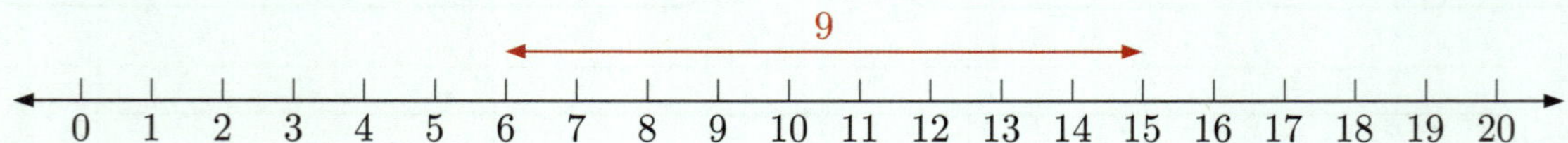

The difference between 6 and 15 is 9.

Exercise 9

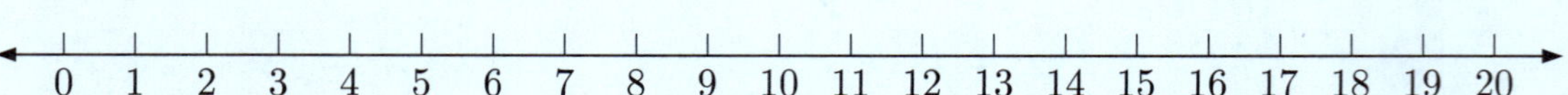

Write down the difference between:

a 5 and 9 _____

b 13 and 5 _____

c 8 and 17 _____

Discussion

lesser ←――――――――――――――――→ greater

To find the *difference* between two numbers, I subtract the _________________ number from the _________________ number.

If two numbers are equal, the difference between them is _________ .

Exercise 10

For each pair of numbers, write a subtraction to find their difference:

a 3 and 7

_______ − _______ = _______

b 10 and 4

_______ − _______ = _______

c 6 and 12

_______ − _______ = _______

d 7 and 16

_______ − _______ = _______

e 18 and 9

_______ − _______ = _______

f 8 and 13

_______ − _______ = _______

Exercise 11

Complete these subtractions.

a $30 - 10 =$ _______

b $50 - 10 =$ _______

c $20 - 10 =$ _______

d $80 - 10 =$ _______

e $60 - 10 =$ _______

f $100 - 10 =$ _______

Exercise 12

Complete these subtractions.

a $26 - 10 =$ _______

b $39 - 10 =$ _______

c $41 - 10 =$ _______

d $64 - 10 =$ _______

e $97 - 10 =$ _______

f $68 - 10 =$ _______

g $85 - 10 =$ _______

h $72 - 10 =$ _______

i $103 - 10 =$ _______

Puzzle

If I roll two ordinary dice, the *difference* between the numbers could be _______ , _______ , _______ , _______ , _______ , or _______ .

There is $\boxed{1}$ way to get the difference $\boxed{5}$. It is: $6 - 1$.

There are $\boxed{}$ ways to get the difference $\boxed{4}$.

They are: _______________________________________ .

There are $\boxed{}$ ways to get the difference $\boxed{}$.

They are: _______________________________________ .

There are $\boxed{}$ ways to get the difference $\boxed{}$.

They are: _______________________________________ .

There are $\boxed{}$ ways to get the difference $\boxed{}$.

They are: _______________________________________

_______________________________________ .

There are $\boxed{}$ ways to get the difference $\boxed{}$.

They are: _______________________________________

_______________________________________ .

Exercise 13

Complete these subtractions.

a $7 - 2 =$ _______

$17 - 2 =$ _______

$27 - 2 =$ _______

$37 - 2 =$ _______

b $5 - 3 =$ _______

$15 - 3 =$ _______

$35 - 3 =$ _______

$65 - 3 =$ _______

c $9 - 6 =$ _______

$29 - 6 =$ _______

$49 - 6 =$ _______

$89 - 6 =$ _______

Exercise 14

Complete these subtractions.

a $65 - 3 =$ _______

b $86 - 3 =$ _______

c $27 - 5 =$ _______

d $48 - 7 =$ _______

e $59 - 6 =$ _______

f $93 - 2 =$ _______

g $37 - 7 =$ _______

h $88 - 4 =$ _______

i $97 - 6 =$ _______

j $77 - 4 =$ _______

k $68 - 5 =$ _______

l $106 - 4 =$ _______

Exercise 15

Complete these subtractions.

Exercise 16

Complete these subtractions.

a $30 - 4 =$ _______

b $70 - 8 =$ _______

c $40 - 9 =$ _______

d $50 - 1 =$ _______

e $60 - 3 =$ _______

f $80 - 5 =$ _______

g $40 - 7 =$ _______

h $90 - 2 =$ _______

i $100 - 8 =$ _______

Exercise 17

Start with these.

$15 - 7 =$ ______

$16 - 7 =$ ______

$17 - 7 =$ ______

$18 - 7 =$ ______

$19 - 7 =$ ______

$20 - 7 =$ ______

$21 - 7 =$ ______

$22 - 7 =$ ______

Start with these.

$16 - 8 =$ ______

$17 - 8 =$ ______

$18 - 8 =$ ______

$19 - 8 =$ ______

$20 - 8 =$ ______

$21 - 8 =$ ______

$22 - 8 =$ ______

$23 - 8 =$ ______

Start with this.

$17 - 9 =$ ______

$18 - 9 =$ ______

$19 - 9 =$ ______

$20 - 9 =$ ______

$21 - 9 =$ ______

$22 - 9 =$ ______

$23 - 9 =$ ______

$24 - 9 =$ ______

Discussion

Some numbers seem more difficult to subtract than others.

With a partner, or as a group, discuss ways to do each subtraction:

$23 - 9 =$ ______ $26 - 8 =$ ______ $31 - 4 =$ ______

Exercise 18

Complete each subtraction:

a $14 - 9 =$ ______

b $16 - 8 =$ ______

c $25 - 9 =$ ______

d $37 - 8 =$ ______

e $35 - 9 =$ ______

f $22 - 8 =$ ______

g $23 - 9 =$ ______

h $24 - 8 =$ ______

i $41 - 8 =$ ______

j $52 - 9 =$ ______

k $45 - 8 =$ ______

l $36 - 8 =$ ______

m $47 - 8 =$ ______

n $43 - 9 =$ ______

o $62 - 9 =$ ______

p $84 - 7 =$ ______

q $63 - 6 =$ ______

r $75 - 7 =$ ______

Exercise 19

Complete these subtractions.

a
$8 - 2 =$ _____
$80 - 20 =$ _____

b
$9 - 5 =$ _____
$90 - 50 =$ _____

c
$6 - 4 =$ _____
$60 - 40 =$ _____

d
$7 - 6 =$ _____
$70 - 60 =$ _____

e
$8 - 5 =$ _____
$80 - 50 =$ _____

f
$9 - 7 =$ _____
$90 - 70 =$ _____

Exercise 20

Complete these subtractions.

a
$60 - 20 =$ _____
$61 - 20 =$ _____
$64 - 20 =$ _____
$66 - 20 =$ _____

b
$90 - 40 =$ _____
$92 - 40 =$ _____
$93 - 40 =$ _____
$97 - 40 =$ _____

c
$80 - 50 =$ _____
$83 - 50 =$ _____
$85 - 50 =$ _____
$86 - 50 =$ _____

d
$50 - 30 =$ _____
$52 - 30 =$ _____
$55 - 30 =$ _____
$58 - 30 =$ _____

e
$80 - 20 =$ _____
$81 - 20 =$ _____
$86 - 20 =$ _____
$89 - 20 =$ _____

f
$70 - 60 =$ _____
$74 - 60 =$ _____
$77 - 60 =$ _____
$79 - 60 =$ _____

Exercise 21

Complete these subtractions.

a
$12 - 5 =$ _____
$120 - 50 =$ _____

b
$15 - 7 =$ _____
$150 - 70 =$ _____

c
$14 - 9 =$ _____
$140 - 90 =$ _____

Exercise 22

a Take away 3 each time:

18 → ☐ → ☐ → ☐ → ☐ → ☐

b Subtract 4 each time:

30 → ☐ → ☐ → ☐ → ☐ → ☐

c Take away 6 each time:

58 → ☐ → ☐ → ☐ → ☐ → ☐

d Subtract 7 each time:

72 → ☐ → ☐ → ☐ → ☐ → ☐

e Take away 9 each time:

80 → ☐ → ☐ → ☐ → ☐ → ☐

f Subtract 8 each time:

104 → ☐ → ☐ → ☐ → ☐ → ☐

Exercise 23

Describe what is happening in each sequence:

a 25 → 21 → 17 → 13 → 9

I start with _______ , and subtract _______ each time.

b 21 → 19 → 17 → 15 → 13

____

c 22 → 19 → 16 → 13 → 10

____

d 46 → 41 → 36 → 31 → 26

____

Find the missing number: $12 - \boxed{} = 5$

If I start with 12, how many should I subtract to leave 5?

$12 - 7 = 5$

Exercise 24

Find the missing number.

a $6 - \boxed{} = 2$

b $10 - \boxed{} = 7$

c $9 - \boxed{} = 6$

d $8 - \boxed{} = 1$

e $11 - \boxed{} = 0$

f $13 - \boxed{} = 8$

g $12 - \boxed{} = 4$

h $16 - \boxed{} = 10$

i $17 - \boxed{} = 9$

j $18 - \boxed{} = 7$

k $21 - \boxed{} = 15$

l $23 - \boxed{} = 16$

Discussion

What methods did you use to answer **Exercise 24**?

Puzzle

| 2 | 4 | 8 | 9 | 10 | 11 | 12 | 15 | 17 |

Use the numbers once each to make *three* subtractions.

$\boxed{} - \boxed{} = \boxed{} \qquad \boxed{} - \boxed{} = \boxed{}$

$\boxed{} - \boxed{} = \boxed{}$

Can you do this so the *total* of the three answers is 23?

Find the missing number: $\boxed{} - 6 = 2$

How many do I need to start with, so that when I subtract 6, I still have 2?

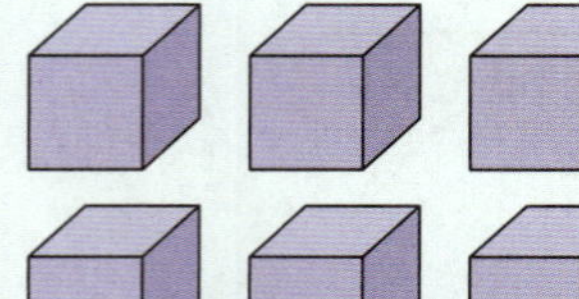

6 taken away   2 left

$8 - 6 = 2$

Exercise 25

Find the missing number.

a $\boxed{} - 7 = 2$ **b** $\boxed{} - 5 = 1$ **c** $\boxed{} - 4 = 9$

d $\boxed{} - 3 = 7$ **e** $\boxed{} - 7 = 5$ **f** $\boxed{} - 8 = 14$

g $\boxed{} - 5 = 16$ **h** $\boxed{} - 8 = 12$ **i** $\boxed{} - 3 = 16$

j $\boxed{} - 2 = 17$ **k** $\boxed{} - 7 = 14$ **l** $\boxed{} - 9 = 17$

Discussion

What methods did you use to answer **Exercise 25**?

Exercise 26

Circle addition words in red, and subtraction words in blue.

altogether   plus   decrease   sum   less

add   subtract   increase   take away   difference

minus   total   how many left

Exercise 27

Circle the words which tell you to add or subtract, then answer the question.

a Lucy has 6 plums and Megan has 5 plums.

How many plums are there altogether?

There are ☐ plums altogether.

b Bruce has 15 biscuits. He gives 6 biscuits to Ella.

How many biscuits does Bruce have left?

Bruce has ☐ biscuits left.

c Sarah puts 5 lollies in a bowl. Her mother adds 7 more lollies.

How many lollies are now in the bowl?

There are now ☐ lollies in the bowl.

d Josh's basket holds 15 eggs. He trips and 9 eggs are smashed.

How many eggs are left?

☐ eggs are left.

e Teagan has 8 blue pens and 3 red pens.

How many pens does she have in total?

In total, Teagan has ☐ pens.

f Lucky was given 12 lettuce leaves.
She ate 8 of them.

How many lettuce leaves does Lucky have left?

Lucky has ☐ lettuce leaves left.

g Veronica has 14 white sheep and 5 black sheep.

How many fewer black sheep does Veronica have than white sheep?

Veronica has ☐ fewer black sheep than white sheep.

Listening Activity

Click to open a Listening Activity.

1 _______________   2 _______________   3 _______________

4 _______________   5 _______________   6 _______________

7 _______________   8 _______________   9 _______________

To subtract numbers with more than one digit, we can use **column subtraction**.

We write the numbers in columns so the place values line up.

We then subtract each column, starting with the units.

$27 - 14$

	T	U
	2	7
−	1	4
	1	3

So, $27 - 14 = 13$.

1 Set out your subtraction.

2 Subtract the units:

$$7 - 4 = 3$$

We write 3 below the line in the units column.

3 Subtract the tens:

$$2 - 1 = 1$$

We write 1 below the line in the tens column.

Exercise 28

Complete these subtractions.

a $27 - 15$

	T	U
	2	7
−	1	5

b $35 - 14$

	T	U
	3	5
−	1	4

c $37 - 12$

	T	U
	3	7
−	1	2

d $37 - 25$

	T	U
	3	7
−	2	5

e $39 - 23$

	T	U
	3	9
−	2	3

f $58 - 34$

	T	U
	5	8
−	3	4

Exercise 29

Set out these subtractions and find the answer.

a 36 − 14

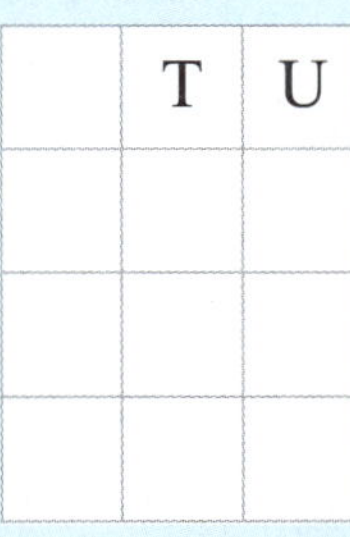

b 45 − 23

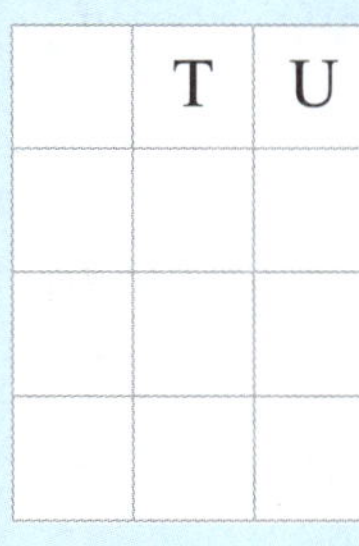

c 48 − 25

d 49 − 34

e 57 − 22

f 68 − 44

Exercise 30

Create a subtraction to find the answer.

a Alyssa has 39 stamps. She uses 24 of them.

How many stamps does Alyssa have left?

Alyssa has ☐ stamps left.

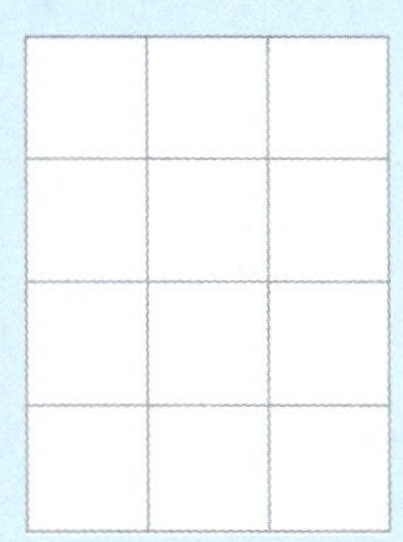

b Yuki has 47 lollies. She gives 25 lollies to Lily.

How many lollies does Yuki have left?

Yuki has ☐ lollies left.

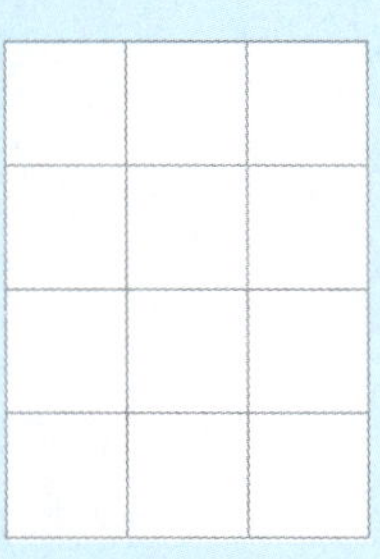

c Mrs Paynter has 88 crayons in her art box.
53 crayons are new.

How many crayons are not new?

Mrs Paynter has ☐ crayons which are not new.

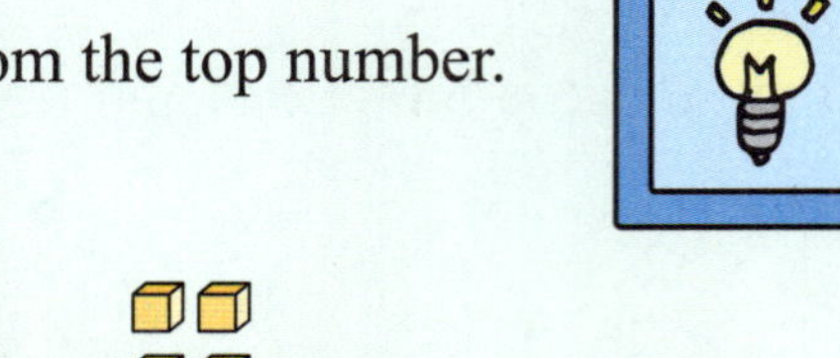

	T	U
	5	3
−	1	6

We *always* begin a subtraction in the units column.

We *always* subtract the bottom number from the top number.

But what do we do with $3 - 6$?

To complete this subtraction we **exchange** 1 ten for 10 units

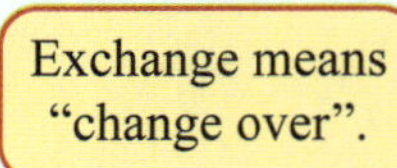

	T	U
	4	13
	5̶	3̶
−	1	6
	3	7

53 is now written as 4 tens and 13 units.

$13 - 6 = 7$ units.

$4 - 1 = 3$ tens.

So, $53 - 16 = 37$.

Exercise 31

Complete these subtractions. You will need to use exchanging.

a $32 - 14$

	T	U
	3	2
−	1	4

b $34 - 16$

	T	U
	3	4
−	1	6

c $41 - 19$

	T	U
	4	1
−	1	9

d $43 - 25$

	T	U
	4	3
−	2	5

e $52 - 27$

	T	U
	5	2
−	2	7

f $33 - 29$

	T	U
	3	3
−	2	9

Exercise 32

Set out these subtractions and find the answer.

a $54 - 35$

b $65 - 28$

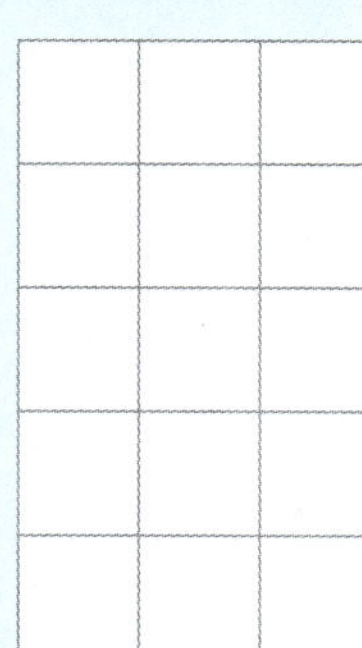

c $71 - 59$

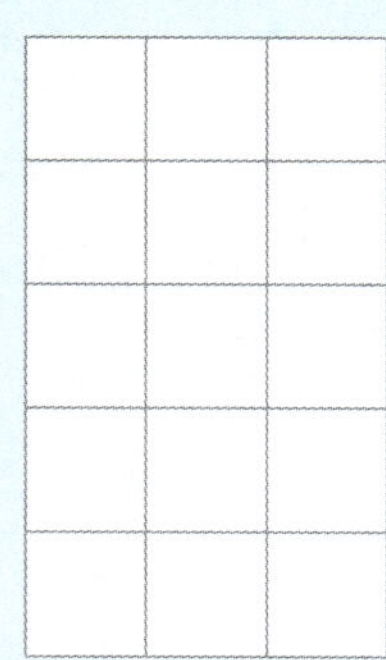

Exercise 33

Create a subtraction to find the answer.

a Adam picked 73 apricots. He gave 39 apricots to his father.

How many apricots did Adam keep for himself?

Adam kept ☐ apricots for himself.

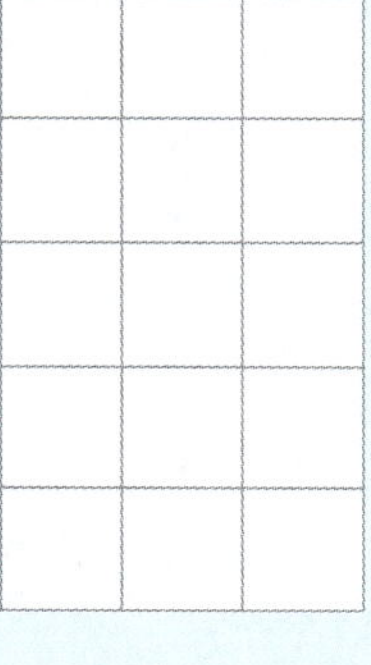

b There are 43 spiders in Brayden's garage. His grandma removes 27 spiders.

How many spiders are left in Brayden's garage?

☐ spiders are left in Brayden's garage.

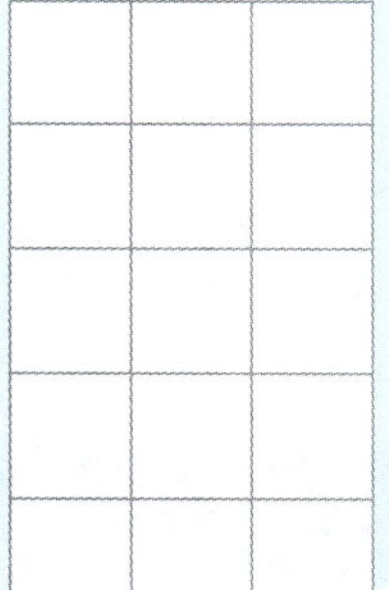

c Mr Telfer has 84 exam papers to check. He has checked 58.

How many exam papers have not been checked?

☐ exam papers have not been checked.

Exercise 34

Solve each problem using addition or subtraction:

a Wendy has 25 stamps. She uses 17 of them.

How many stamps does Wendy have left?

Wendy has ☐ stamps left.

b Ruslan collects red and blue toy cars. He has 17 red cars and 19 blue cars.

How many cars does Ruslan have altogether?

Ruslan has ☐ cars in his collection.

c Sam grew 16 pumpkins and 33 eggplants.

- How many vegetables did Sam grow altogether?

  Sam grew ☐ vegetables altogether.

- How many fewer pumpkins than eggplants did Sam grow?

  Sam grew ☐ fewer pumpkins than eggplants.

Puzzle

Charlotte has 62 chocolate eggs.

She gave 17 chocolate eggs to Deanne.

How many more chocolate eggs does Charlotte now have than Deanne?

Charlotte now has ☐ more chocolate eggs than Deanne.

Revision

1 Use the number line to help you subtract:

a 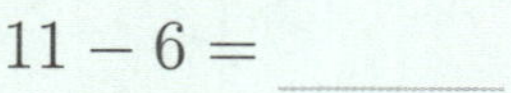
$11 - 6 = $ _______

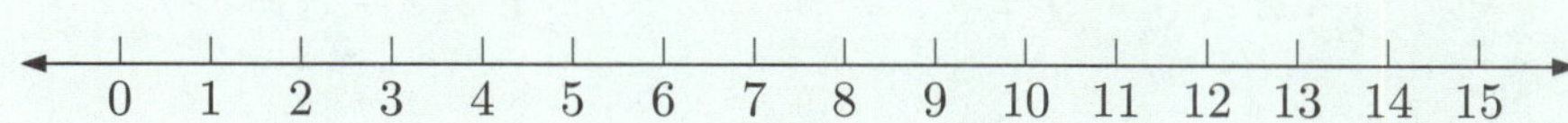

b 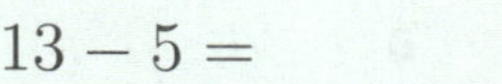
$13 - 5 = $ _______

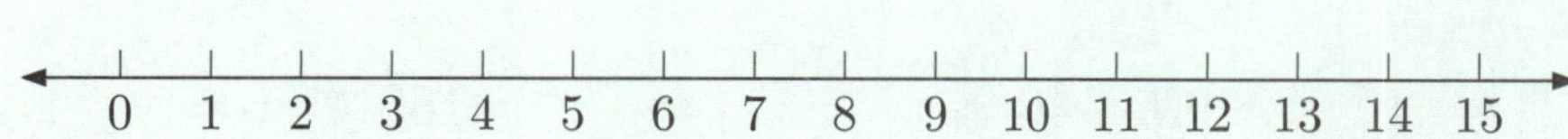

2 Use the addition to write *two* subtractions.

a $\qquad 7 + 9 = 16$ **b** $\qquad 5 + 8 = $ _______

so $16 - $ _______ $= $ _______   so   _______ $- 5 = $ _______

and $16 - $ _______ $= $ _______   and   _______ $- 8 = $ _______

3 Write down the result of each subtraction:

a $8 - 4 = $ _______   **b** $7 - 6 = $ _______   **c** $9 - 3 = $ _______

d $10 - 6 = $ _______   **e** $14 - 8 = $ _______   **f** $17 - 7 = $ _______

g $15 - 7 = $ _______   **h** $18 - 9 = $ _______   **i** $13 - 6 = $ _______

4 Complete:

a 7 minus $4 = $ _______   **b** 9 take away $5 = $ _______   **c** 6 subtract $2 = $ _______

d Subtract 8 from 12:   _______ $- $ _______ $= $ _______

5 For each pair of numbers, write a subtraction to find their difference:

a 12 and 3   **b** 8 and 17

_______ $- $ _______ $= $ _______   _______ $- $ _______ $= $ _______

6 Complete each subtraction:

a $70 - 10 = $ _______   **b** $48 - 10 = $ _______   **c** $37 - 10 = $ _______

7 Complete each subtraction:

a $39 - 4 = $ _______   **b** $46 - 5 = $ _______   **c** $78 - 6 = $ _______

d $60 - 2 = $ _______   **e** $40 - 4 = $ _______   **f** $50 - 8 = $ _______

8 Complete each subtraction:

a $62 - 7 = $ _______

b $43 - 8 = $ _______

c $54 - 9 = $ _______

d $71 - 6 = $ _______

e $80 - 9 = $ _______

f $65 - 8 = $ _______

9 Complete these subtractions.

a $7 - 3 = $ _______

$70 - 30 = $ _______

b $9 - 4 = $ _______

$90 - 40 = $ _______

c $16 - 8 = $ _______

$160 - 80 = $ _______

10 Complete these subtractions.

a

$30 - 20 = $ _______

$31 - 20 = $ _______

$34 - 20 = $ _______

$36 - 20 = $ _______

$38 - 20 = $ _______

b

$60 - 30 = $ _______

$63 - 30 = $ _______

$65 - 30 = $ _______

$67 - 30 = $ _______

$69 - 30 = $ _______

11 a Take away 4 each time:

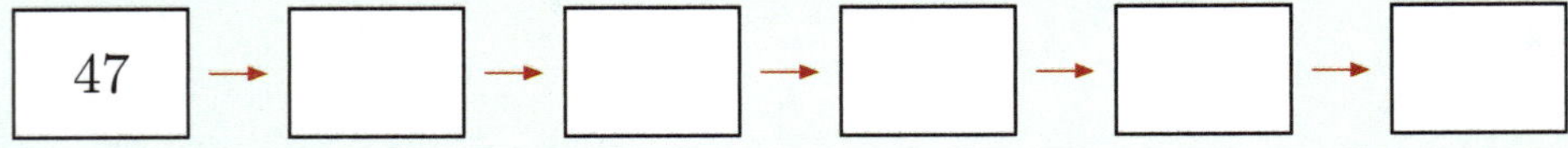

b Take away 7 each time:

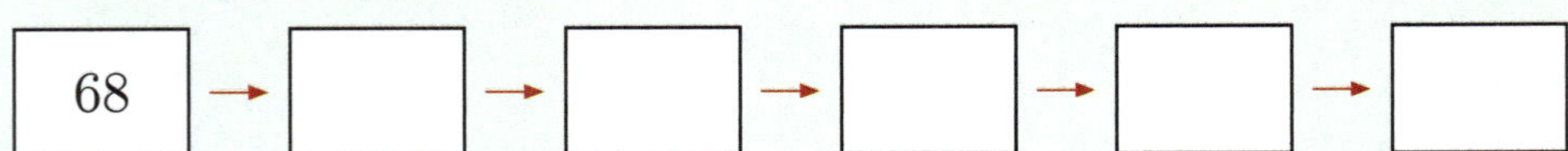

12 Describe what is happening in the sequence.

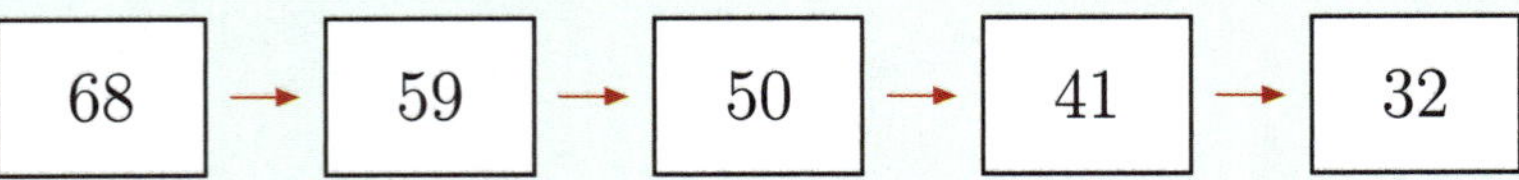

I start with _______ , and subtract _______ each time.

13 Fill in the missing number.

a $9 - \boxed{} = 1$

b $12 - \boxed{} = 7$

c $19 - \boxed{} = 13$

d $16 - \boxed{} = 8$

e $13 - \boxed{} = 3$

f $22 - \boxed{} = 15$

14 Fill in the missing number.

a $\boxed{} - 6 = 3$ **b** $\boxed{} - 2 = 9$ **c** $\boxed{} - 6 = 4$

d $\boxed{} - 10 = 12$ **e** $\boxed{} - 9 = 18$ **f** $\boxed{} - 5 = 16$

15 Circle the words which tell you to subtract, then answer the question.

a Jeremy had 14 eggs. He used 6 eggs in his cooking.

How many eggs does Jeremy have left?

Jeremy has $\boxed{}$ eggs left.

b Carla started the year with 15 baby teeth. Now she has only 8.

How many baby teeth has Carla lost this year?

Carla has lost $\boxed{}$ baby teeth this year.

16 Complete these subtractions.

a

	T	U
	7	5
−	4	1

b

	T	U
	6	8
−	2	4

c

	T	U
	9	7
−	7	2

17 Set out these subtractions and find the answer.

a $38 - 16$ **b** $52 - 31$ **c** $69 - 47$

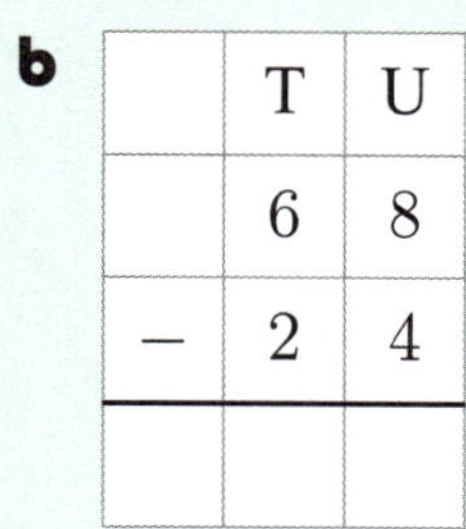

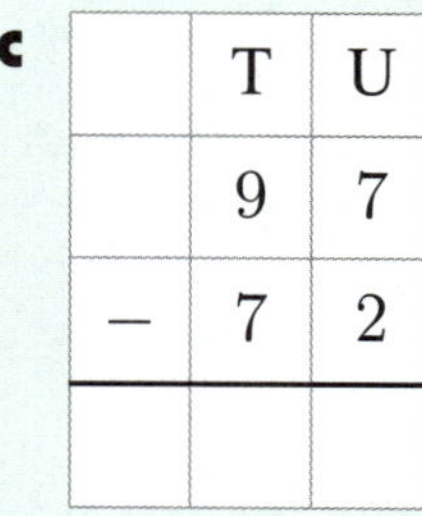

18 Complete these subtractions.

a

	T	U
	9	1
−	3	8

b

	T	U
	8	5
−	4	7

c

	T	U
	8	0
−	5	9

19 Set out these subtractions and find the answer.

   a $73 - 59$ **b** $53 - 17$ **c** $94 - 29$

	T	U

20 Create a subtraction to find the answer.

   a Sally scored 39 goals in a netball game.
Her friend Georgina scored 16 goals.

How many more goals did Sally score than Georgina?

Sally scored [ ] more goals than Georgina.

   b Rico picked 52 strawberries from his garden.
He gave 26 strawberries to Suzanne.

How many strawberries did Rico keep?

Rico kept [ ] strawberries.

CHAPTER 6: LENGTH

A **length** or **distance** is a measurement of how far it is between two points.

We use many different words to describe lengths.

Copy each word into the correct group or groups!

tall	thick	shallow	fat
deep	long	wide	thin
narrow	slender	short	skinny

Length words	*Height words*

Width words	*Depth words*

If you are unsure of the meaning of any of these words, ask your teacher!

A **desk ruler** is used to measure short lengths.

The ruler is a *number line*. The distance from 0 to 1 on the number line is the **unit** of measurement.

For a desk ruler, the unit is a special length called a **centimetre**.

We often write centimetre as **cm**.

How long is this highlighter?

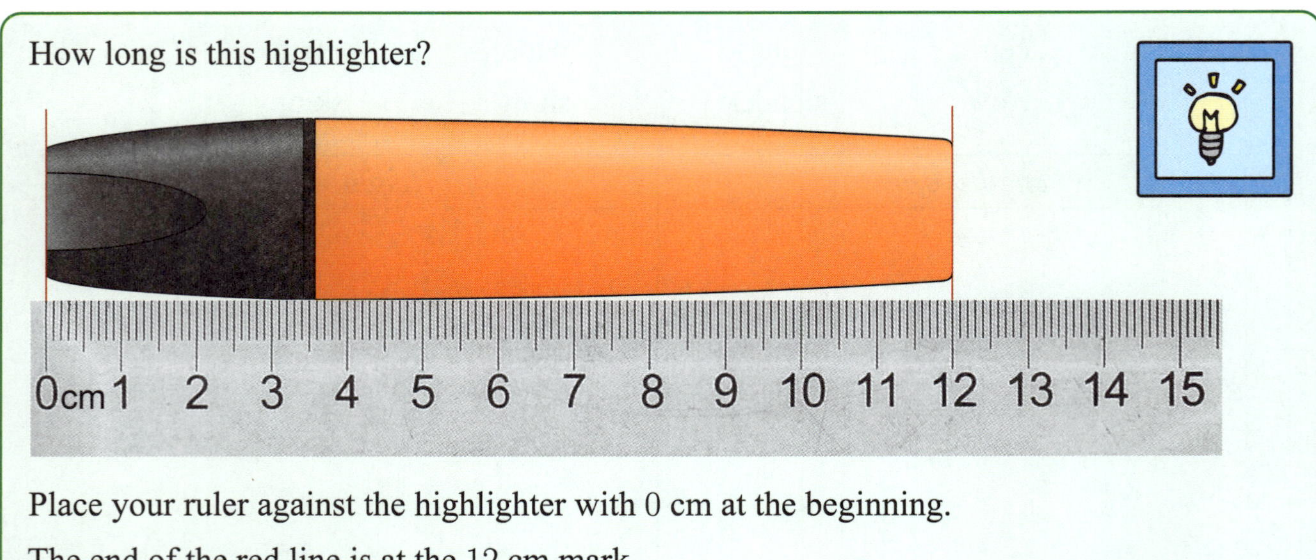

Place your ruler against the highlighter with 0 cm at the beginning.

The end of the red line is at the 12 cm mark.

The highlighter is 12 cm long.

Exercise 1

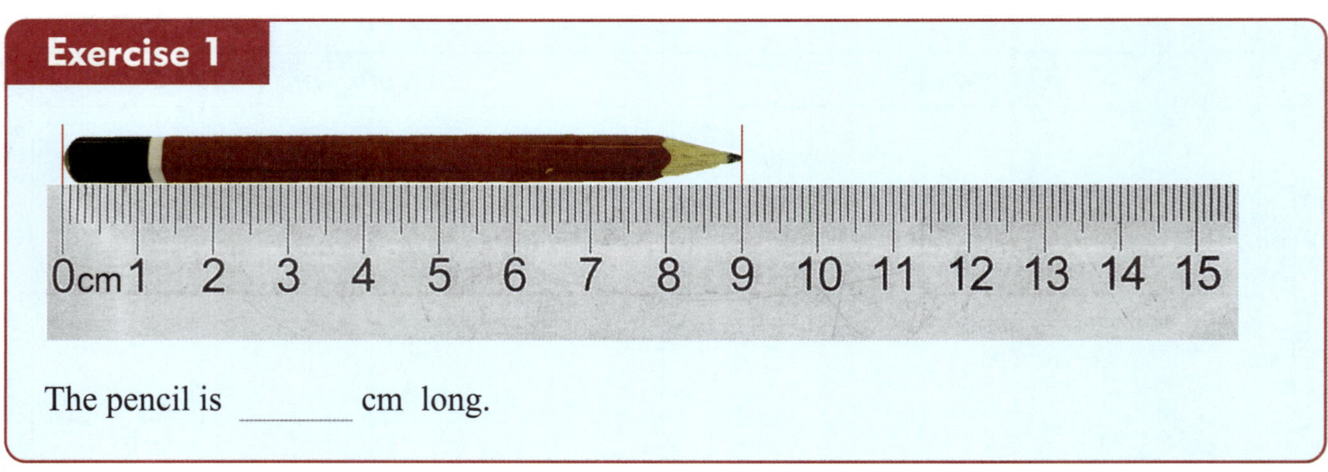

The pencil is _______ cm long.

Discussion

Why is it important to write the units with every length measurement?

Exercise 2

Measure each line carefully. Write your answer in the space provided.

a ________________________________ __________ cm

b ____________ __________ cm

c ____ __________ cm

d ____________________________ __________ cm

e ________________________ __________ cm

f ____ __________ cm

g ____ __________ cm

h ____ __________ cm

i ______________ __________ cm

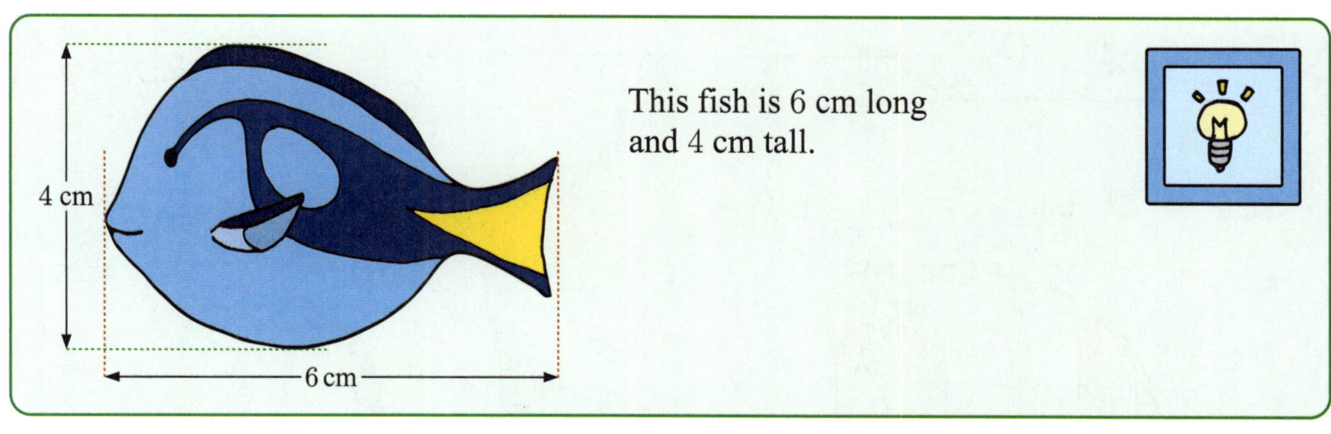

This fish is 6 cm long
and 4 cm tall.

Exercise 3

Measure each length carefully.

The rectangle is __________ cm wide

and __________ cm high.

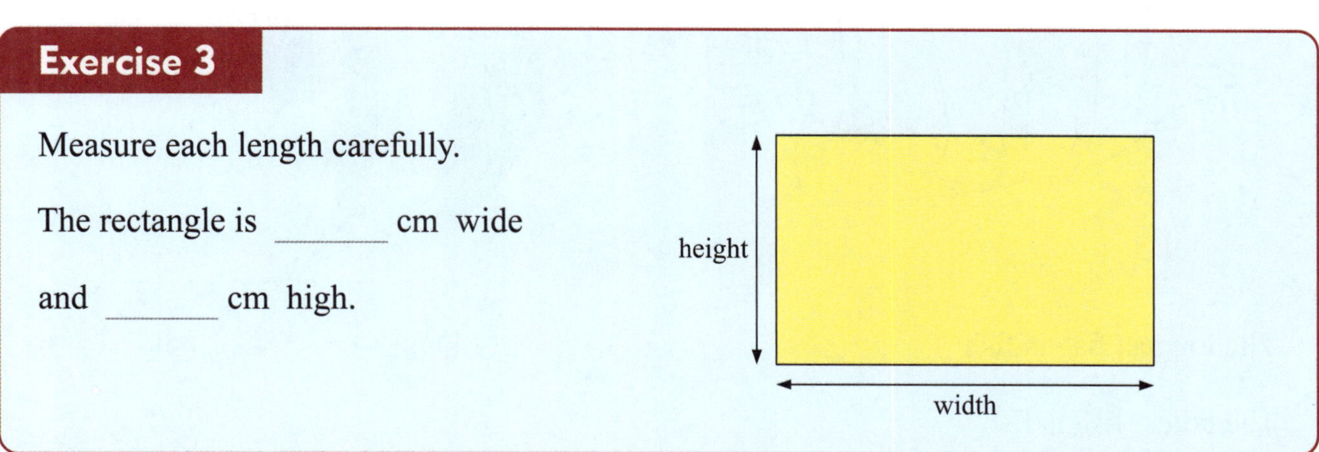

Exercise 4

Measure each fish:

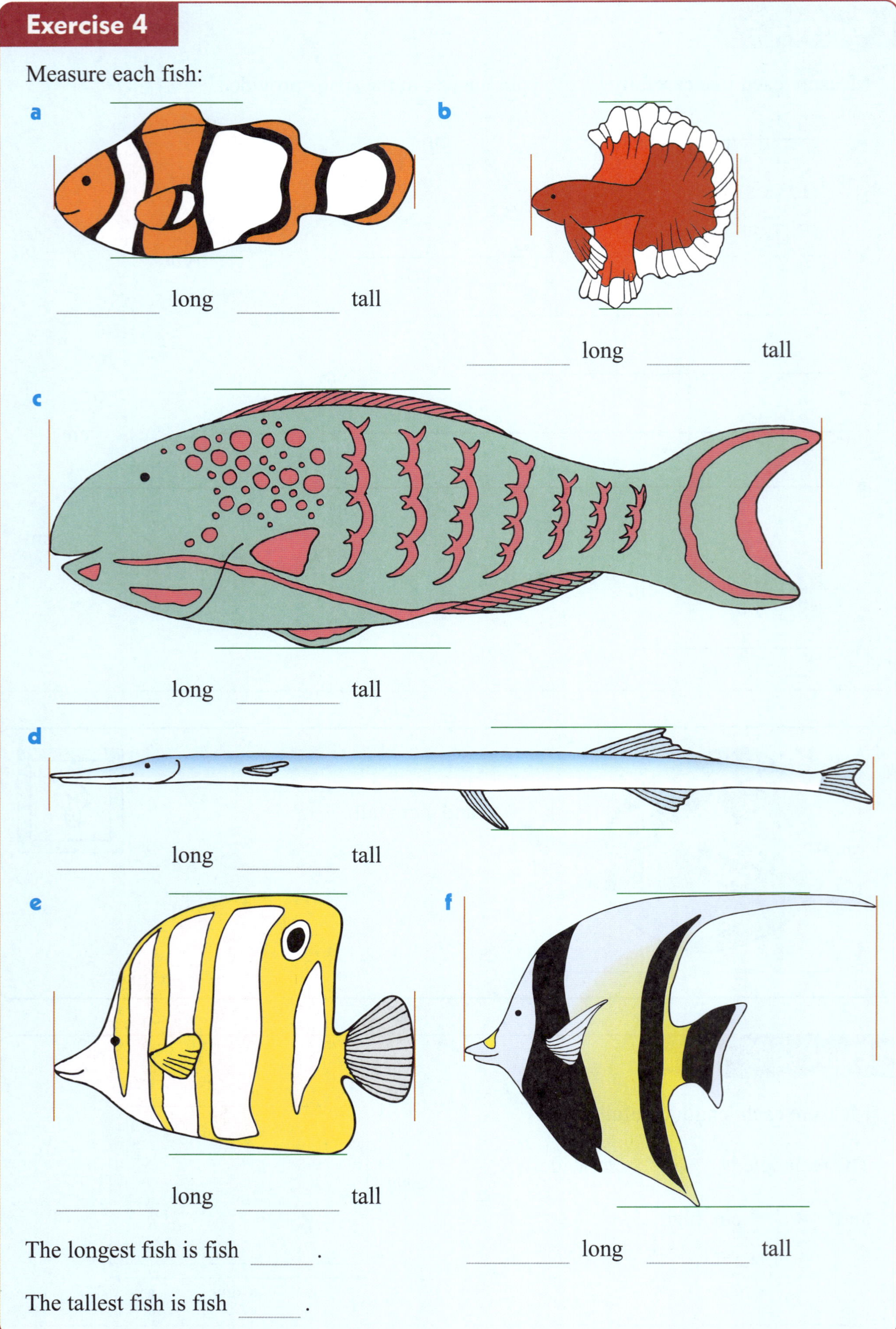

a

__________ long   __________ tall

b

__________ long   __________ tall

c

__________ long   __________ tall

d

__________ long   __________ tall

e

__________ long   __________ tall

f

__________ long   __________ tall

The longest fish is fish __________ .

The tallest fish is fish __________ .

Discussion

Measure the length of this pen.

Is it an exact whole number of centimetres long?

What whole number is its length closest to?

The pen is *about* __________ cm long.

Exercise 5

Measure each length:

a   The paper clip is about __________ cm long.

b   My phone is about __________ cm long and __________ cm wide.

c   The sticky tape holder is about __________ cm long and __________ cm high.

d 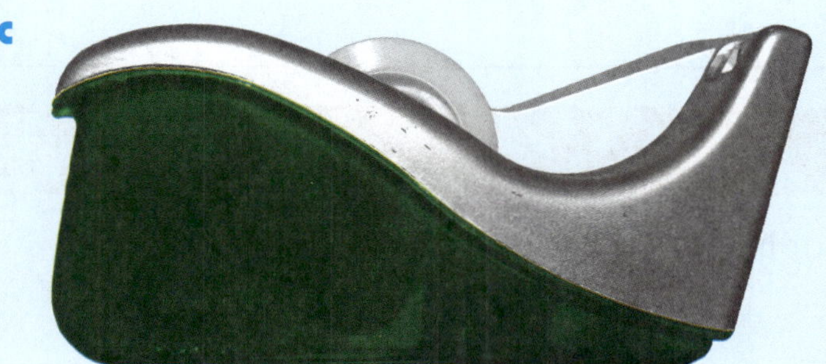

The name badge is about __________ cm long and __________ cm high.

Activity   Comparing lengths

1 Identify 2 objects in your classroom which are longer than this book.

____

____

2 This book is about _________ cm long

and _________ cm wide.

3 Identify 2 objects in your classroom which are shorter than this book.

____

____

4 The shortest of these objects is about _________ cm long.

5 Choose 3 objects in your classroom. Draw each object, showing which is the shortest and which is the widest.

Activity

Arrange a group of 5 classmates in order of height, without using a ruler.

tallest

shortest

To measure longer distances, we use a different *unit* called a **metre**.

We often write metre as just **m**.

$$1 \text{ metre} = 100 \text{ centimetres}$$
$$1 \text{ m} = 100 \text{ cm}$$

Your teacher might have a **metre ruler** which is 100 cm or 1 m long.

To measure distances in metres, we can use a **tape measure**.

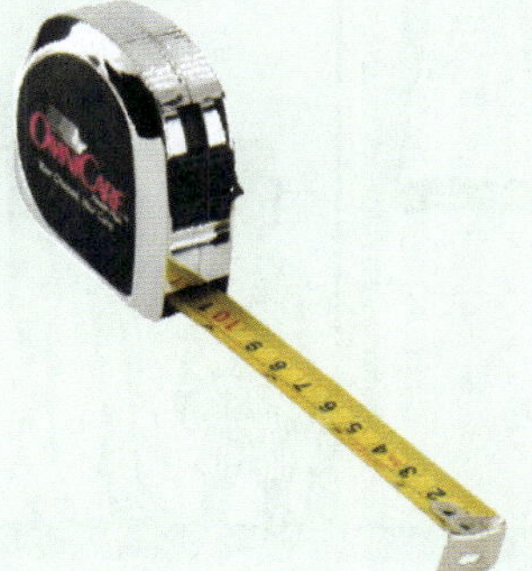

tape measure

reel tape measure

Exercise 6

Write whether you would measure the length of each object in centimetres or metres.

a

b

c

d

e

f

Exercise 7

1 metre = 100 cm

2 metres = _______ cm

3 metres = _______ cm

_______ metres = 400 cm

_______ metres = 500 cm

6 metres = _______ cm

7 metres = _______ cm

_______ metres = 800 cm

_______ metres = 900 cm

10 metres = _______ cm

Discussion

What point is *halfway* along a metre ruler?

How many *centimetres* are there in *half* a metre? _______ cm

Exercise 8

1 m 45 cm =   100 cm   +   45 cm   =   145 cm

a 1 m 80 cm =   100 cm   + _______ cm = _______ cm

b 2 m 65 cm = _______ cm +   65 cm   = _______ cm

c 3 m 15 cm =   300 cm   + _______ cm = _______ cm

d 4 m 20 cm = _______ cm + _______ cm = _______ cm

e 3 m 35 cm = _______ cm + _______ cm = _______ cm

Exercise 9

110 cm =   100 cm   +   10 cm   = 1 m 10 cm

a 120 cm =   100 cm   + __________ cm = 1 m __________ cm

b 175 cm = __________ cm +   75 cm   = __________ m 75 cm

c 235 cm =   200 cm   + __________ cm = __________ m __________ cm

d 390 cm = __________ cm +   90 cm   = __________ m __________ cm

e 285 cm =   200 cm   + __________ cm = __________ m __________ cm

f 440 cm = __________ cm + __________ cm = __________ m 40 cm

g 515 cm = __________ cm +   15 cm   = __________ m __________ cm

Activity

You will need: tape measures, metre ruler

A box is 3-dimensional.

Its *dimensions* are length, width, and height.

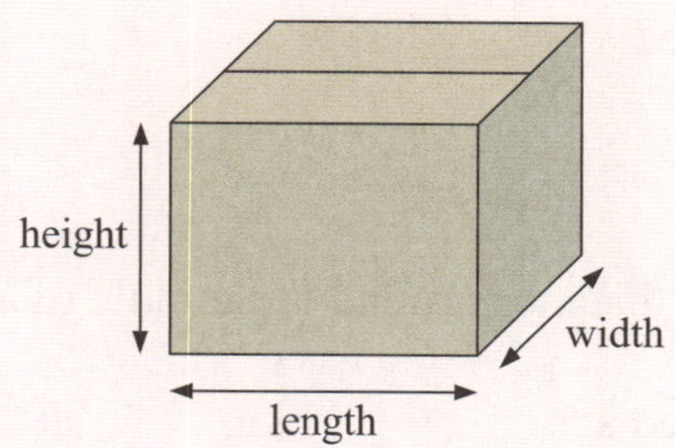

What to do:

Your teacher will give you two boxes labelled **A** and **B**.

Measure their dimensions using a tape measure. Check your answers using the metre ruler.

Write their dimensions on these diagrams.

A

B

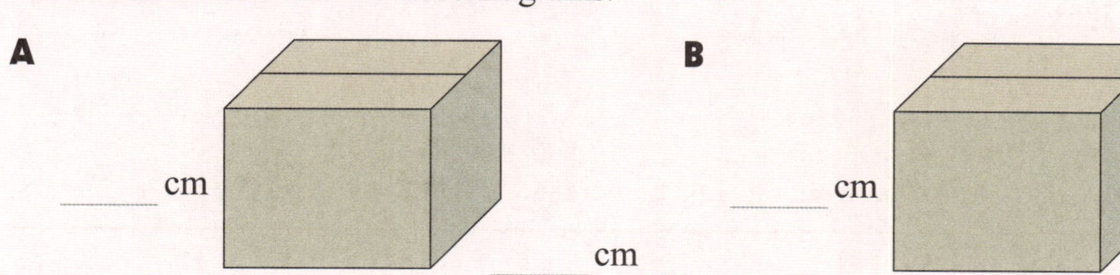

Activity

desk ruler   metre ruler   tape measure   reel tape measure

What to do:

Choose a tool to measure each length. Try more than one tool if you need to.

1 The length of the room is ________________ .

   I used a ________________________________ .

2 The length of my class desk is ______________ .

   I used a ________________________________ .

3 The length of my pencil is ______________ .

   I used a ________________________________ because

   ____ .

4 The distance around the middle of a football is ______________ .

   I used a ________________________________ because

   ____ .

5 The height of the classroom door is ______________ .

   I measured it by ________________________________

   ____

   ____ .

Revision

1 Measure each line carefully. Write your answer in the space provided.

a ___________ cm

b ___________ cm

2 Measure each beetle:

a

___________ long

___________ tall

b

___________ long

___________ tall

c

___________ long

___________ tall

The longest beetle is beetle ______.

The tallest beetle is beetle ______.

3 Measure each length:

a The bolt is about ______ cm long.

b The screw is about ______ cm long.

c The power outlet is about ______ cm long and ______ cm high.

4 Write whether you would measure the height of each object in centimetres or metres.

a

b

____

____

5 Complete:

1 m = 100 cm

______ m = 300 cm

5 m = ______ cm

______ m = 700 cm

10 m = ______ cm

6 Complete:

a 125 cm =   100 cm   + ______ cm = 1 m ______ cm

b 240 cm = ______ cm +   40 cm   = ______ m 40 cm

c 385 cm = ______ cm + ______ cm = ______ m ______ cm

7 Complete:

a 1 m 30 cm = ______ cm +   30 cm   = ______ cm

b 2 m 95 cm =   200 cm   + ______ cm = ______ cm

c 2 m 10 cm = ______ cm + ______ cm = ______ cm

The **capacity** of a container is the amount of liquid it can hold.

Exercise 1

Look at these containers.

baby bottle 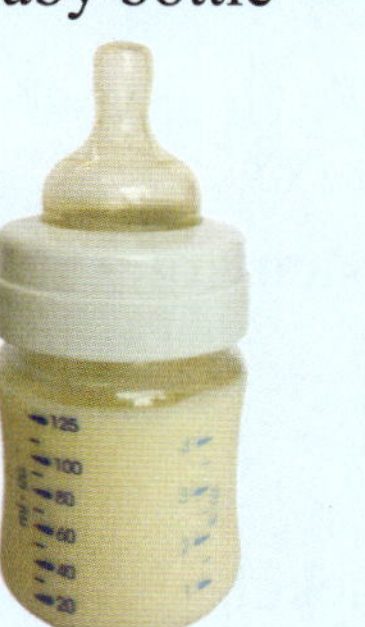

medicine cup 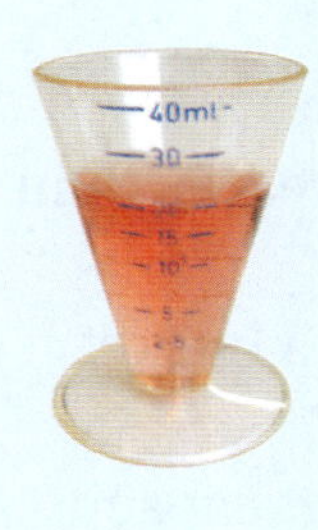

teaspoon 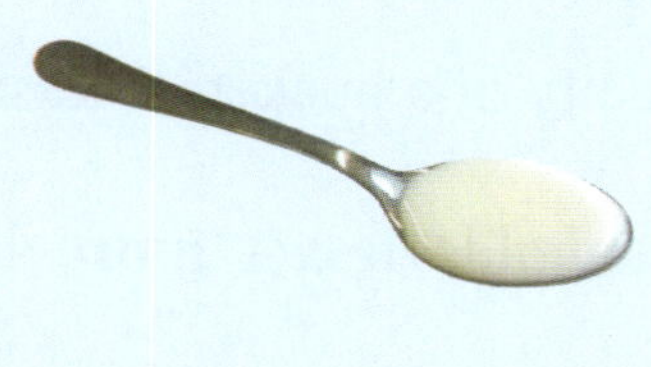

a Which container has the smallest capacity? _______________________

b Which container has the largest capacity? _______________________

Exercise 2

Match each container with the word that best describes its size:

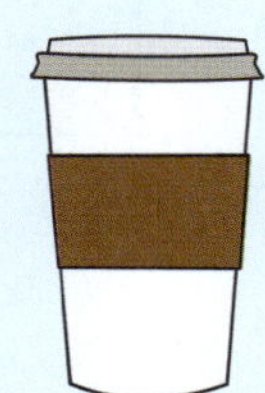

| large | tiny | enormous | small |

Exercise 3

bathtub

bucket

swimming pool

Circle the correct words.

a The capacity of a bucket is **less than** / **greater than** the capacity of a swimming pool.

b A bathtub holds **less** / **more** liquid than a bucket.

c The container with the **smallest** / **largest** capacity is the bucket.

Exercise 4

Arrange these containers in order.

aquarium

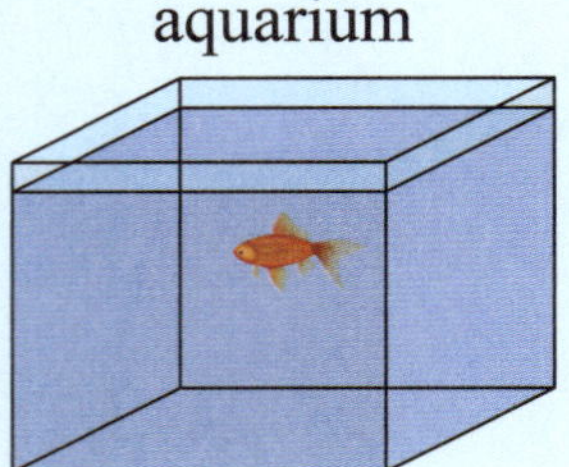

drink bottle

saucepan

least capacity ———————————————————————▶ most capacity

We measure capacity using **litres** (L) or **millilitres** (mL).

$$1 \text{ L} = 1000 \text{ mL}$$

We buy milk in 1 or 2 litre bottles.

A teaspoon holds about 5 mL of liquid.

Exercise 5

Practise your spelling.

litre   millilitre

_______________________   _______________________

_______________________   _______________________

Exercise 6

Complete:

We can write litre as _________ and millilitre as _________ .

Exercise 7

1 L = 1000 mL

2 L = [ ] mL

3 L = [ ] mL

[ ] L = 4000 mL

5 L = [ ] mL

[ ] L = 6000 mL

Activity

List 4 containers you can find in your classroom that hold:

a less than 1 L

b more than 1 L.

We also use litres and millilitres to measure the *amount* of liquid in a container.

We use a **scale** to read the amount of liquid.

A **scale** is a number line used for measuring.

Discussion

Look at this container.

What *units* does it measure in? ____________________

How much liquid fits between each mark on the scale?

____

How much liquid is in the container? ____________________

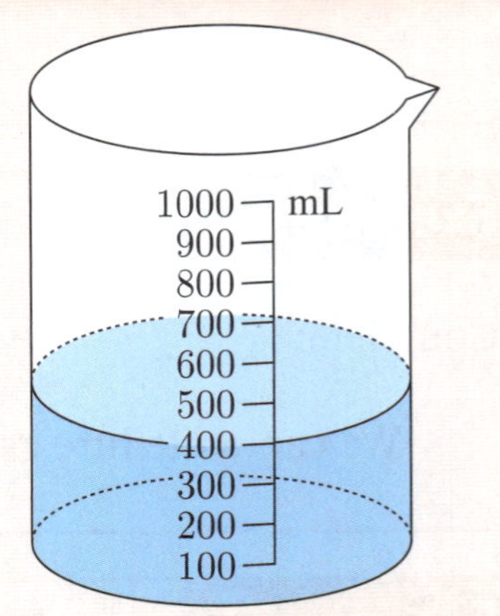

Exercise 8

How much liquid is in the container? Remember to use *units*.

a
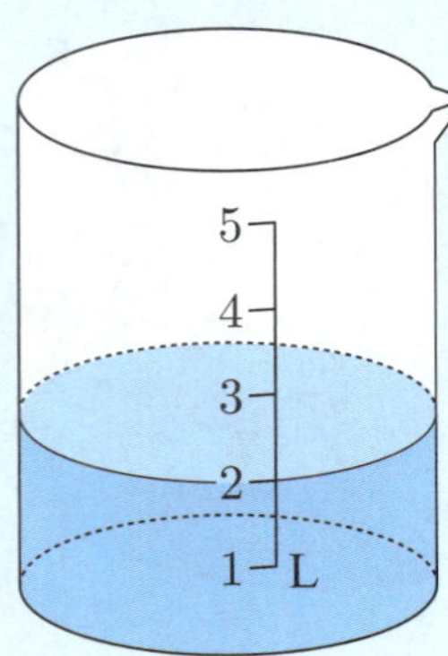

b
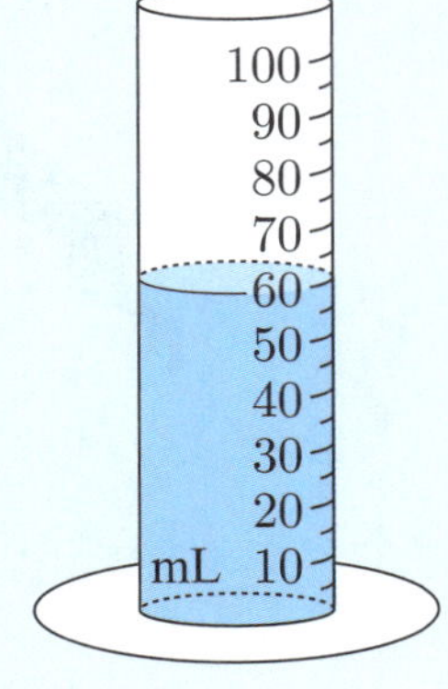

c
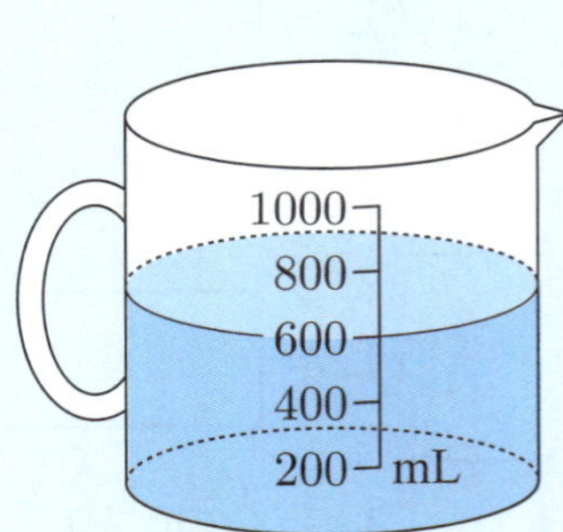

____________________   ____________________   ____________________

d
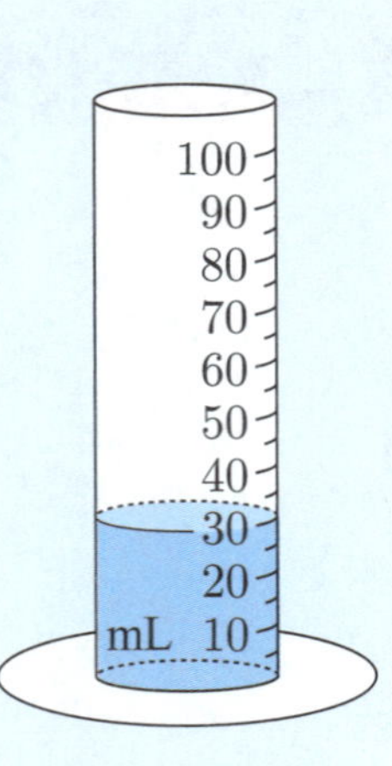

e
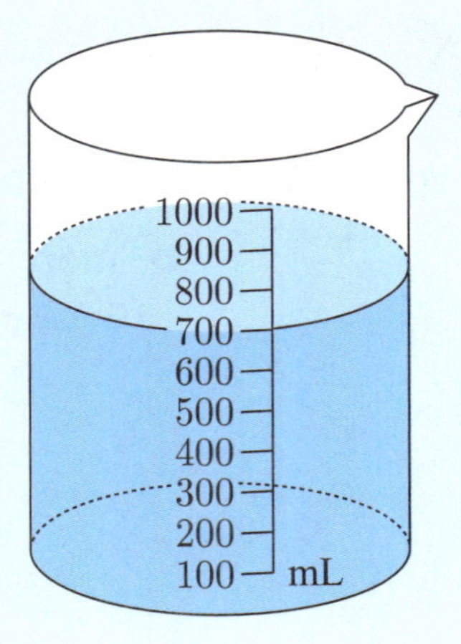

f
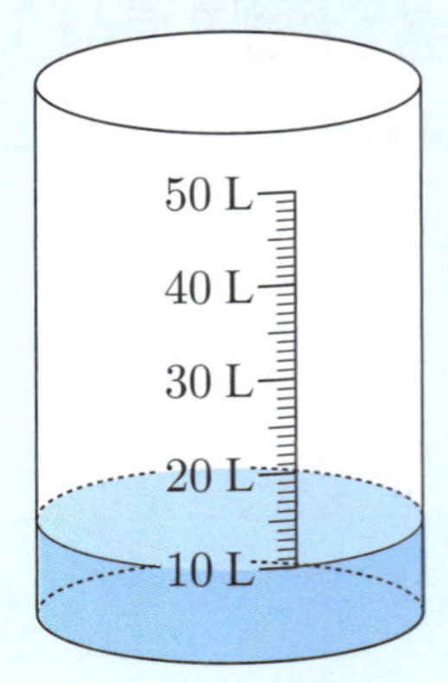

____________________   ____________________   ____________________

Exercise 9

Draw the amount of liquid in the container.

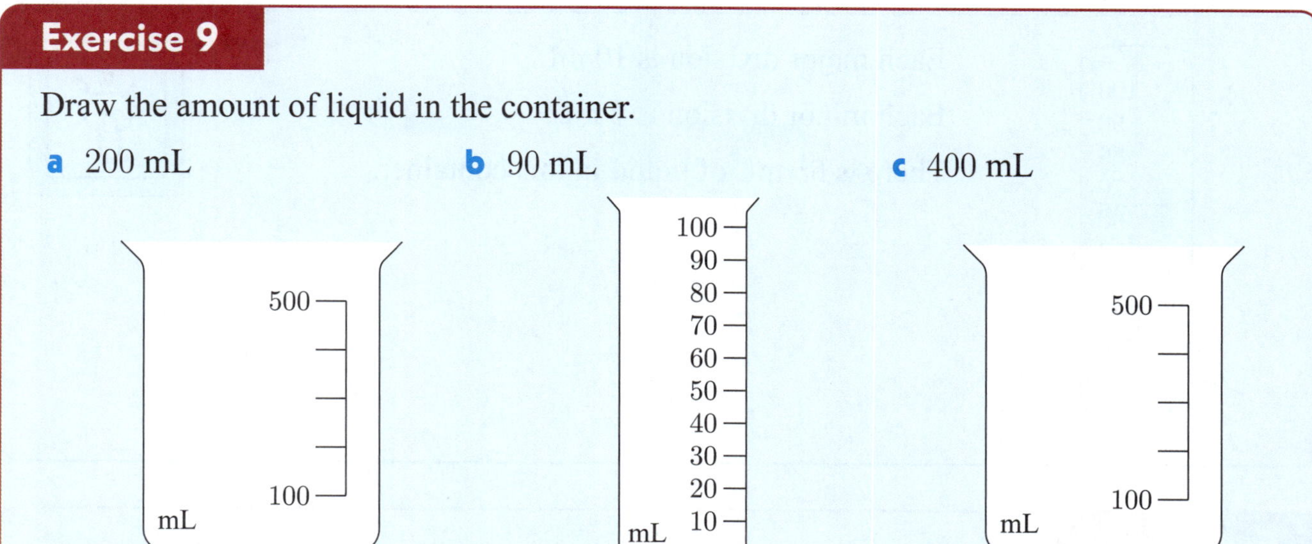

a 200 mL   **b** 90 mL   **c** 400 mL

Exercise 10

How much liquid is in the container?

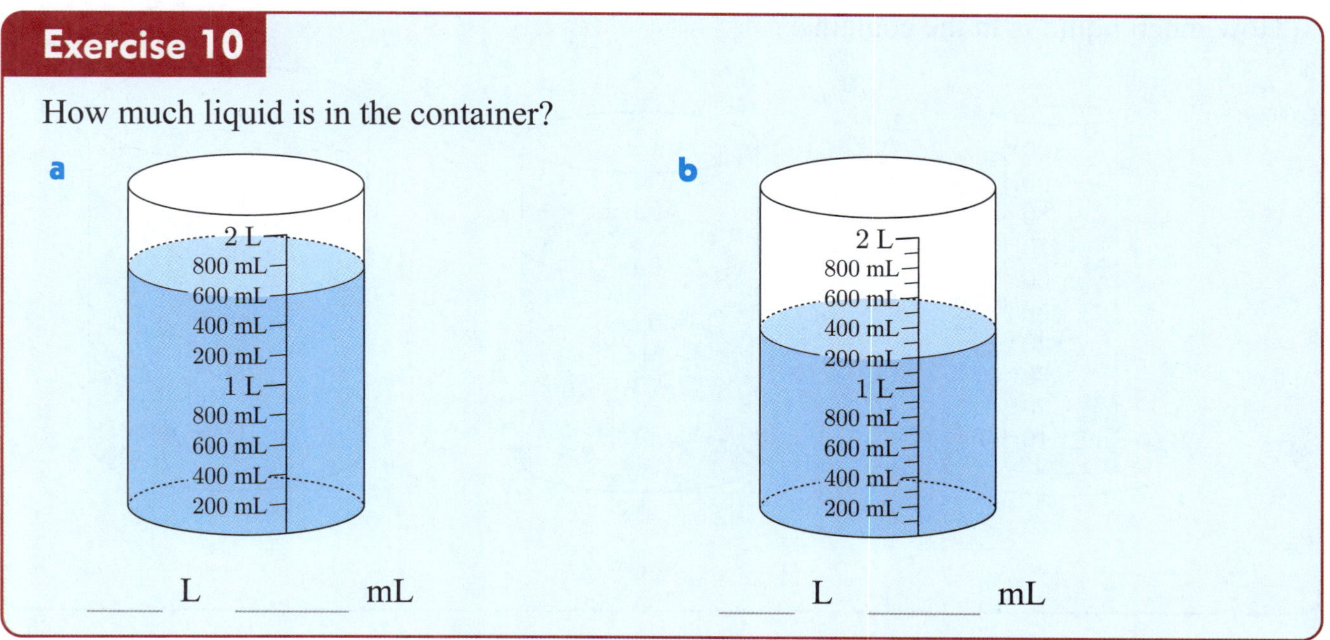

a _______ L _______ mL   **b** _______ L _______ mL

Activity

Your teacher will place 3 labelled containers on their desk.

Draw the liquid in the container and write down the amount.

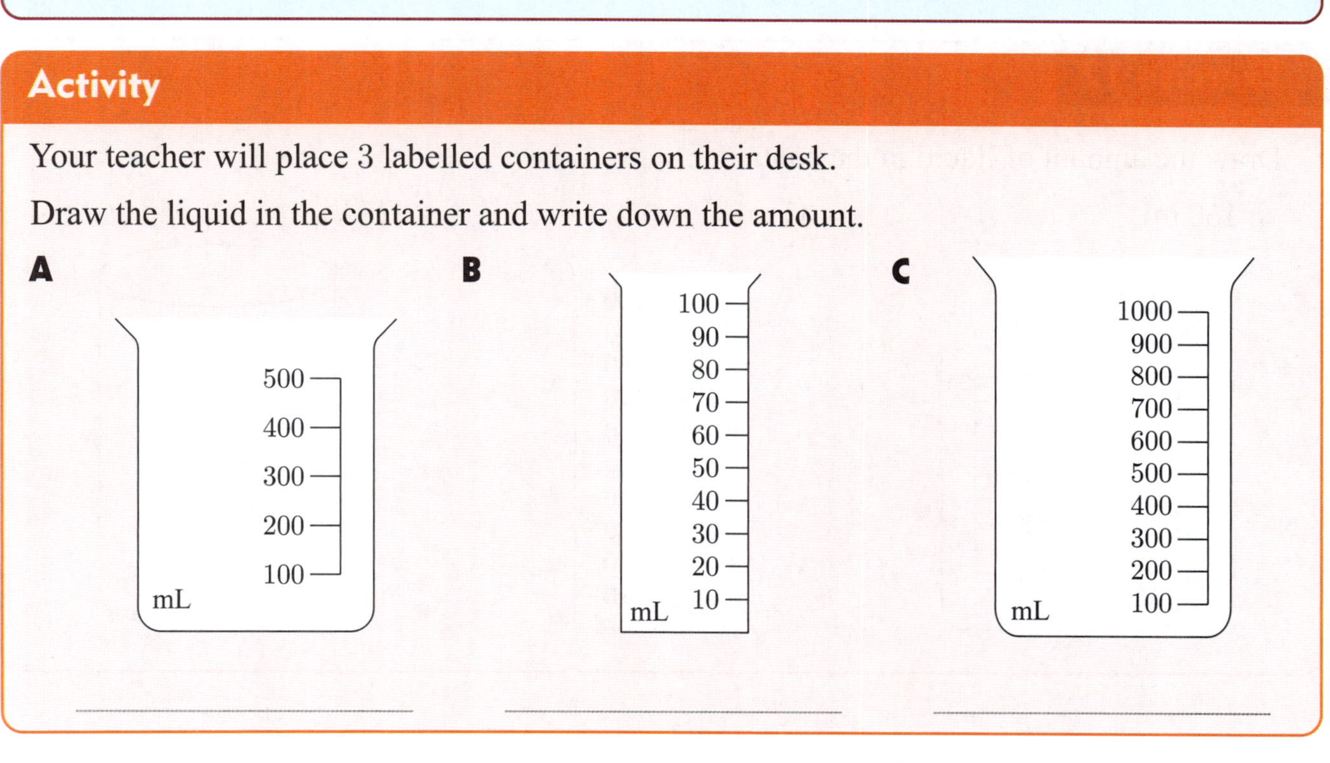

A _______________   **B** _______________   **C** _______________

Each major division is 10 mL.

Each minor division is 5 mL.

There is 65 mL of liquid in this container.

Exercise 11

How much liquid is in the container?

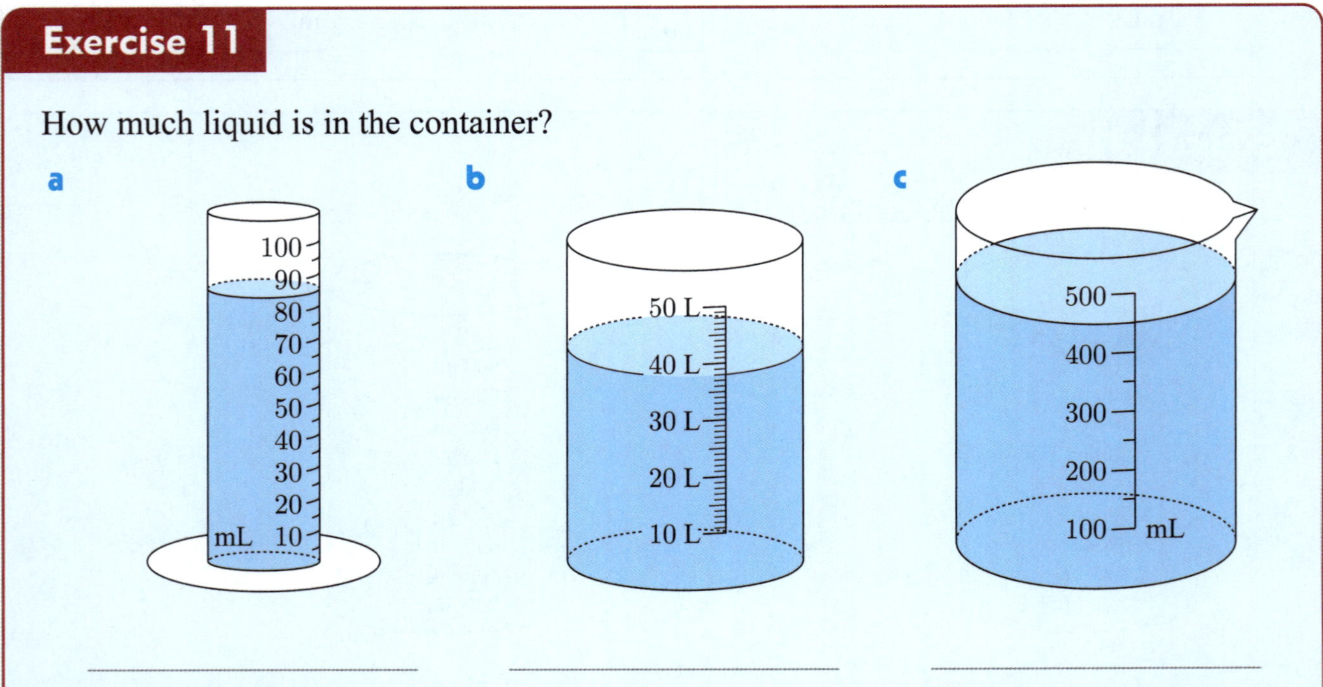

a

b

c

_______________   _______________   _______________

Exercise 12

Draw the amount of liquid in the container.

a 150 mL

b 35 mL

c 27 L

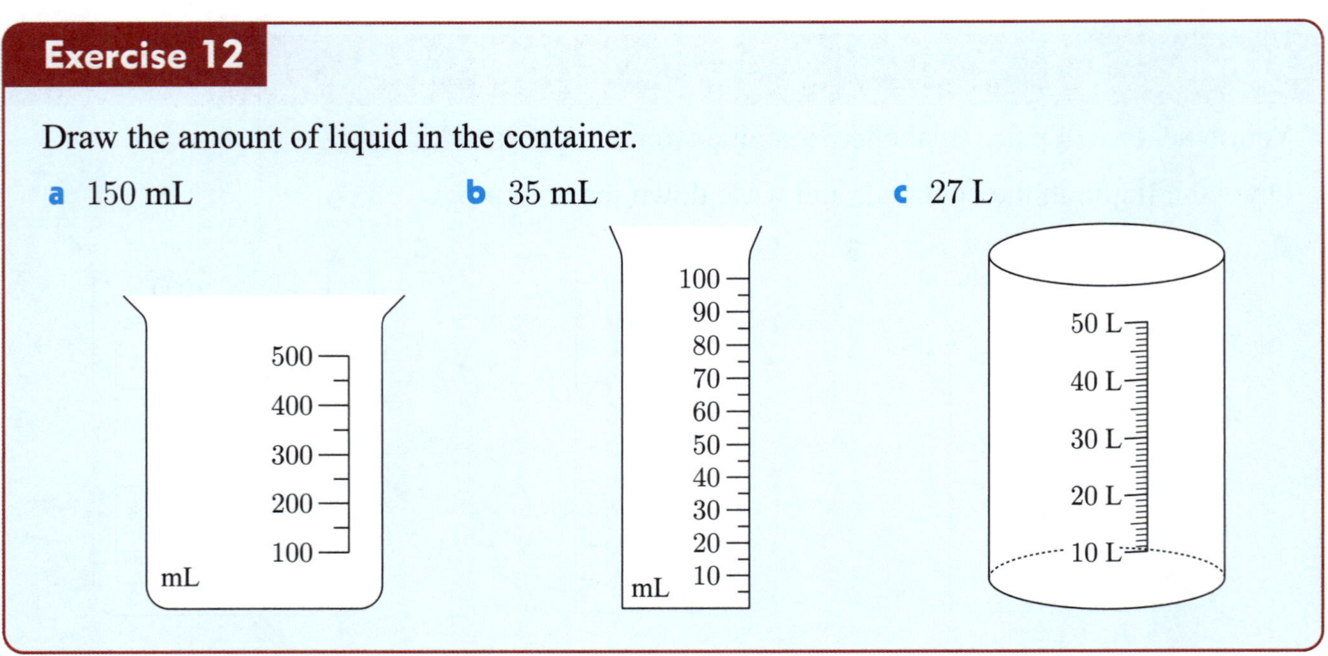

This container contains more than 4 L but less than 5 L of liquid.

Exercise 13

Describe the amount of liquid in each container:

a

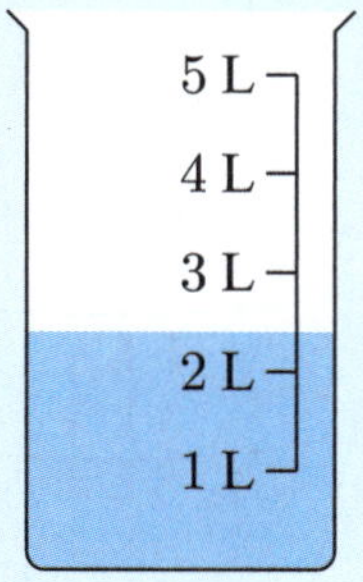

more than __________ but

less than __________

b

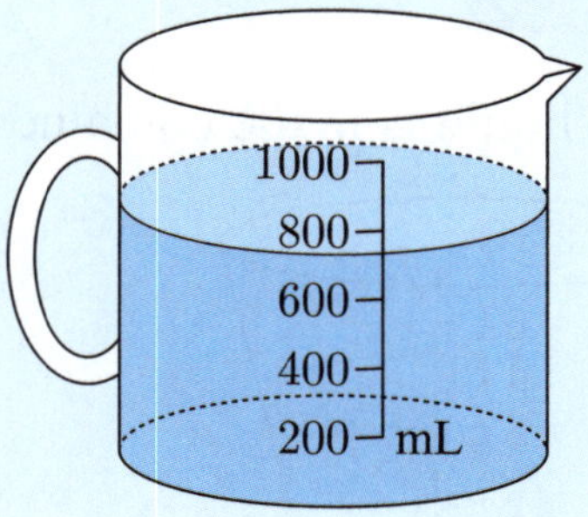

more than __________ but

less than __________

c

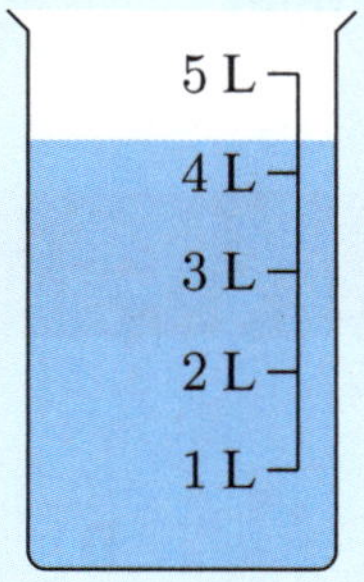

more than __________ but

less than __________

d

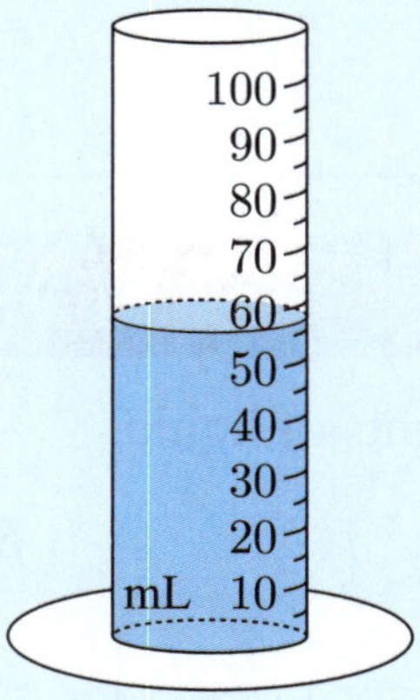

more than __________ but

less than __________

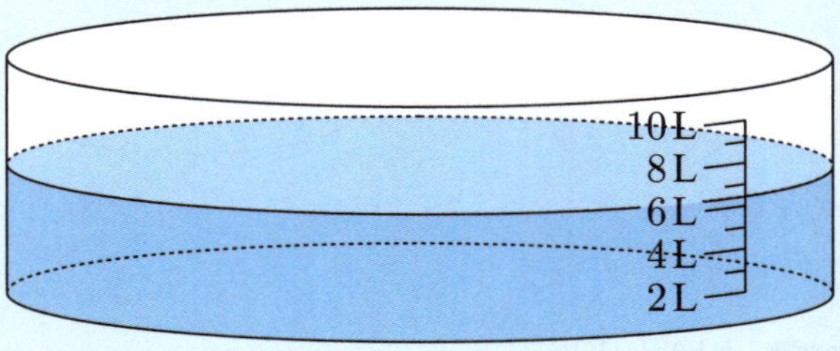

more than __________ but

less than __________

e

Discussion

Why do these scale marks get closer together higher up this bucket?

How much liquid is in the container?   ___________

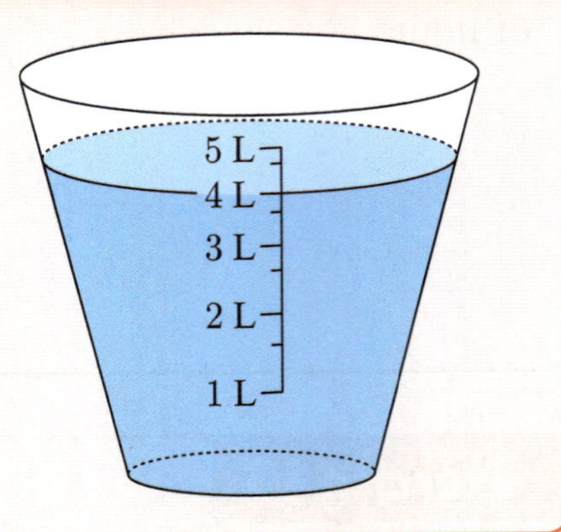

Exercise 14

How much liquid is in the container?

a
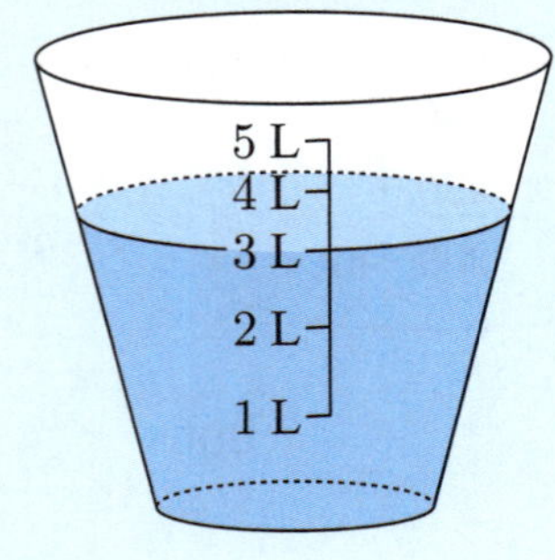

____

b
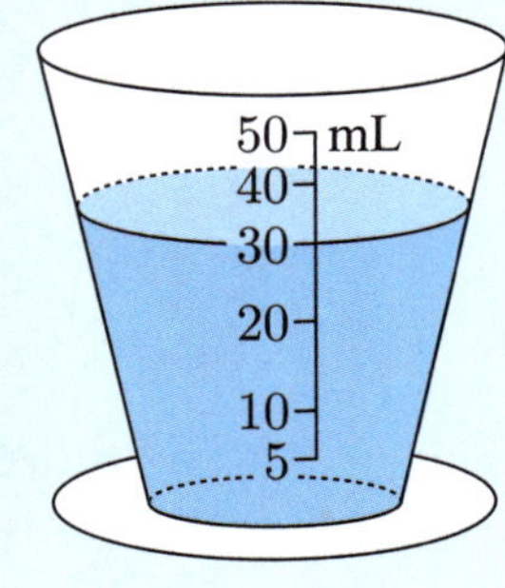

____

c
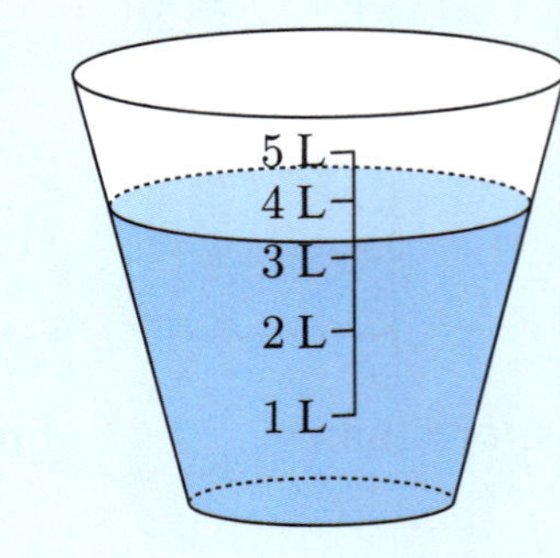

more than ___________ but

less than ___________

Activity

Compare the amounts of liquid.

A
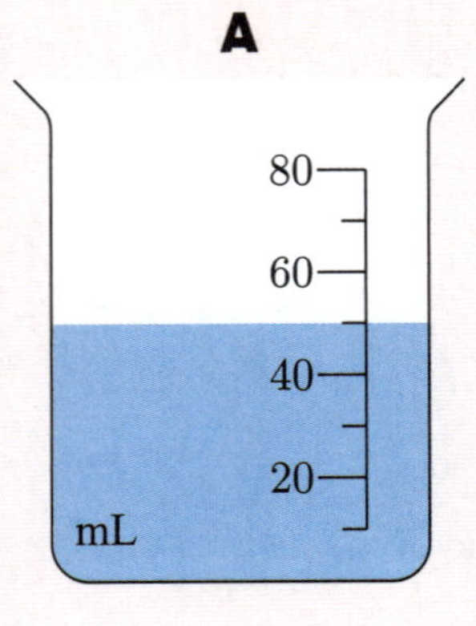

B
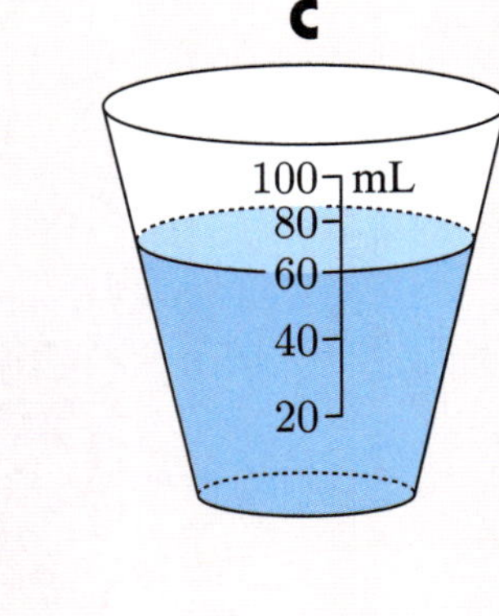

C

D
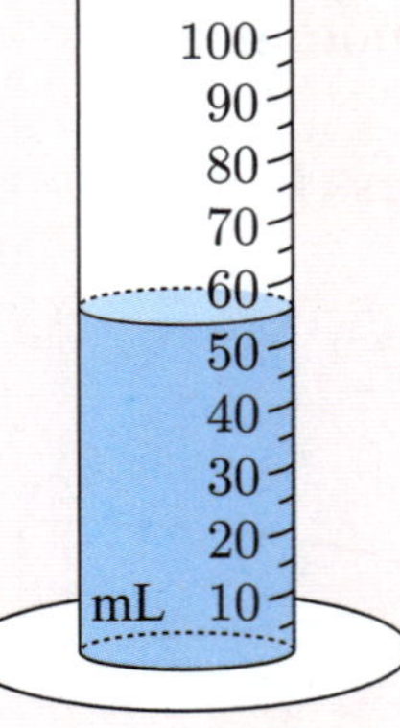

___________ ___________ ___________ ___________

least liquid ———→ most liquid

Revision

1 Look at these containers.

The container with the smallest capacity is __________ .

The container with the largest capacity is __________ .

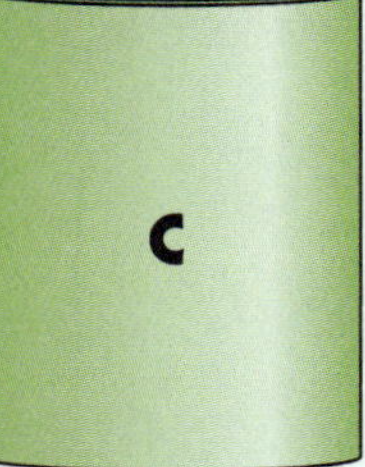

2 Arrange these containers in order.

drinking glass

bin

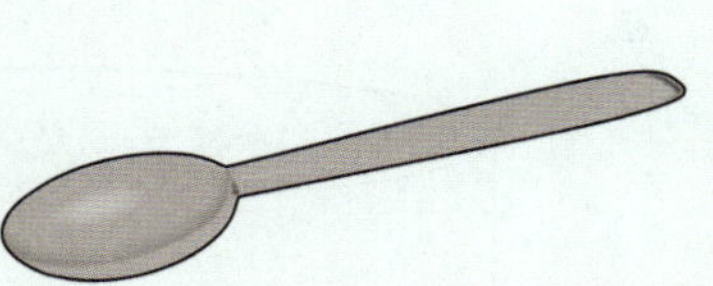

soup spoon

least capacity ⟶ most capacity

3 Complete:

$1 \text{ L} = \boxed{} \text{ mL}$

$\boxed{} \text{ L} = 3000 \text{ mL}$

$5 \text{ L} = \boxed{} \text{ mL}$

$\boxed{} \text{ L} = 7000 \text{ mL}$

4 How much liquid is in the container?

a

b

c

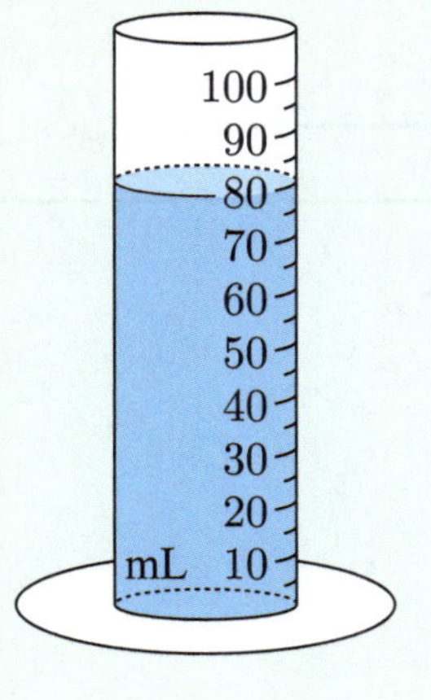

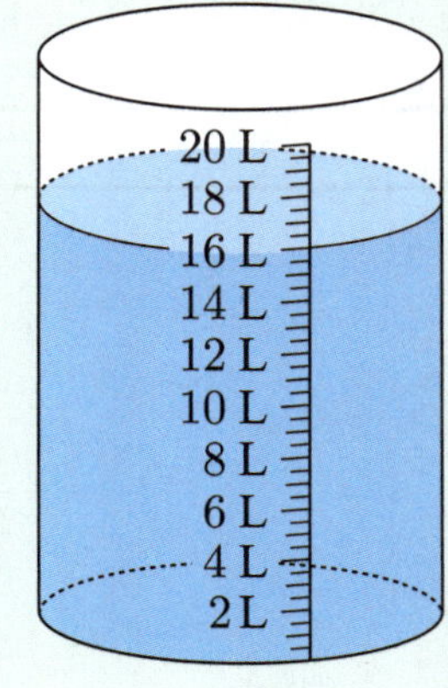

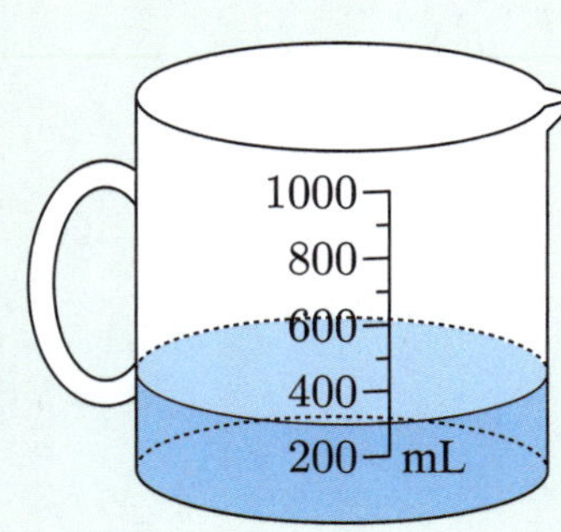

5 Draw the amount of liquid in the container.

a 300 mL

b 250 mL

c 45 mL

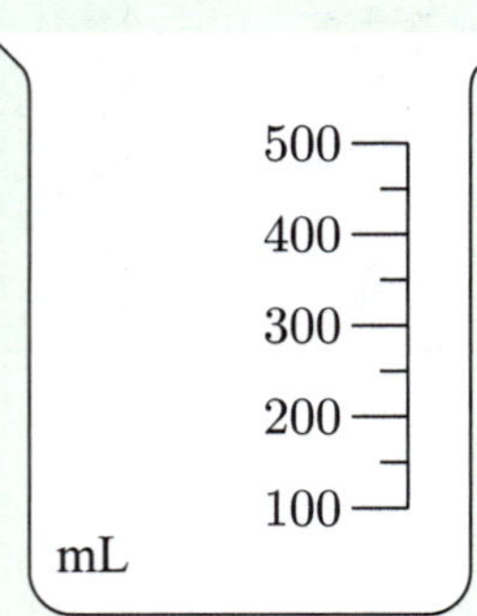 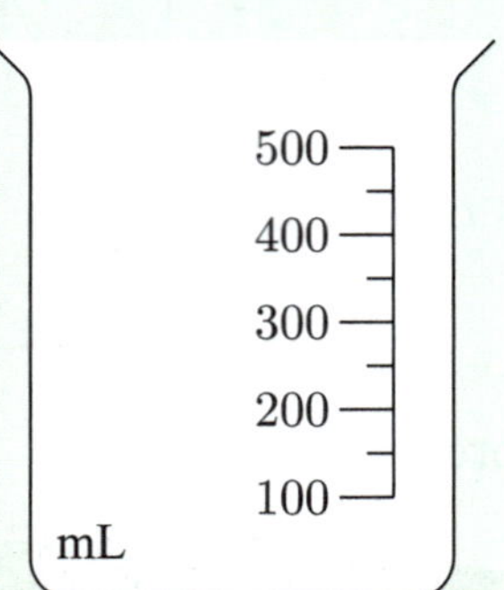 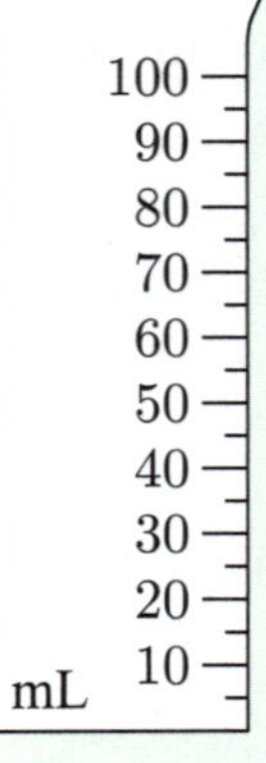

6 Describe the amount of liquid in each container:

a

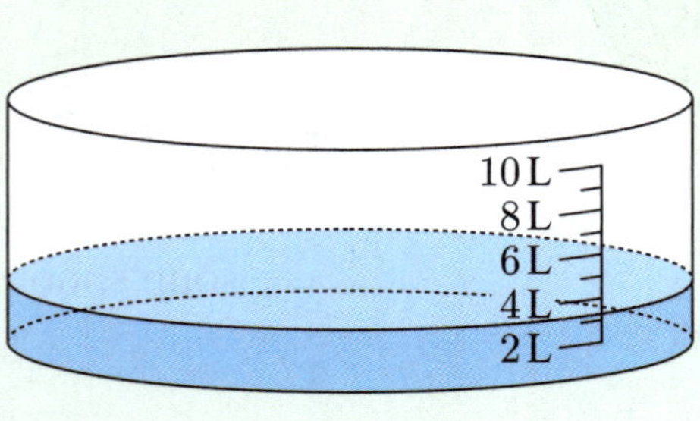

more than _______ but

less than _______

b

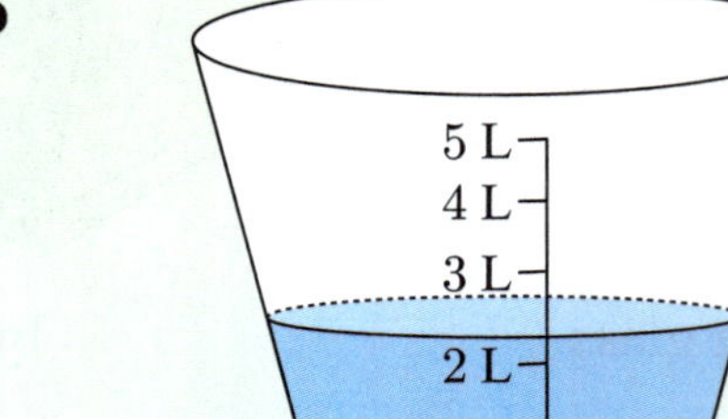

more than _______ but

less than _______

7 Write the amounts of liquid in order.

A 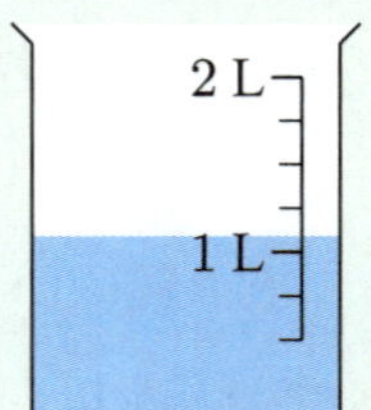 **B** 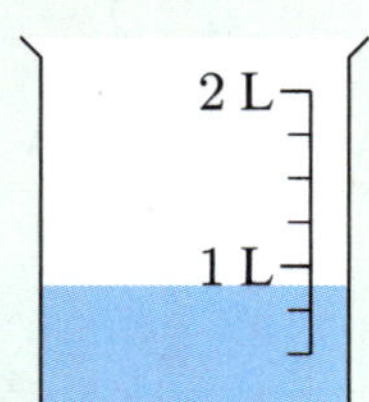 **C** 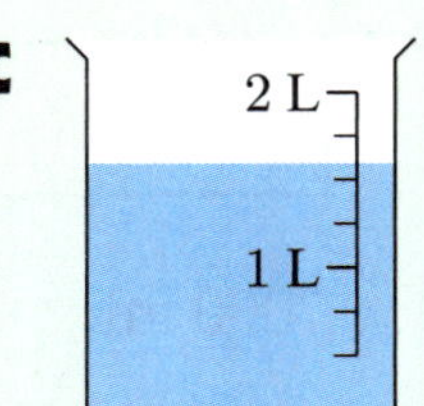 **D** 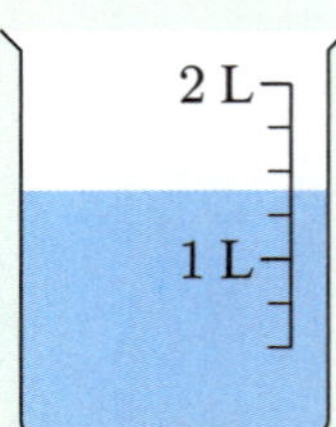

_______ _______ _______ _______

smallest amount ⟶ largest amount

The **mass** of an object is a measure of how *heavy* it is.

When we compare the mass of objects we use the words *heavier* and *lighter*.

To compare mass, we can use a **balance scale**.

The pumpkin is heavier than the egg.

Exercise 1

Use *heavier* or *lighter* to complete each sentence:

a

The lemon is

____

than the book.

b

The cup is

____

than the sugar cube.

c

The cardboard box is

____

than the kettle.

d

The puppy is

____

than the canary.

Puzzle

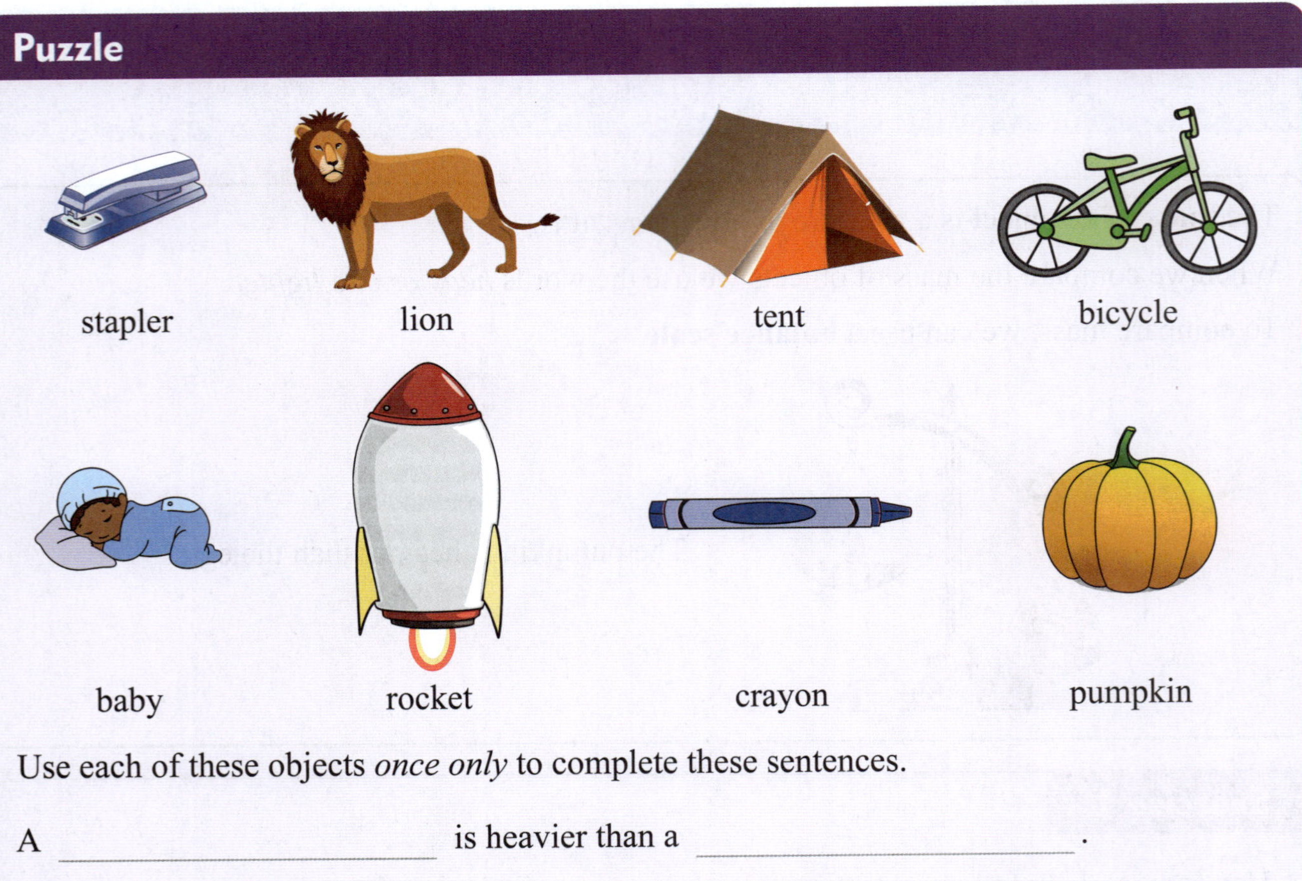

Use each of these objects *once only* to complete these sentences.

A ________________ is heavier than a ________________ .

A ________________ is lighter than a ________________ .

A ________________ is heavier than a ________________ .

A ________________ is lighter than a ________________ .

We measure mass using **grams** (g) or **kilograms** (kg).

$$1 \text{ kg} = 1000 \text{ g}$$

 1 litre of water has mass 1 kilogram.

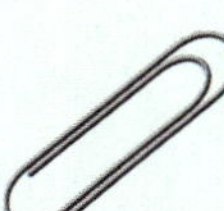 The mass of a paper clip is about 1 gram.

Exercise 2

Complete:

1 kg = 1000 g	☐ kg = 5000 g
2 kg = ☐ g	6 kg = ☐ g
3 kg = ☐ g	7 kg = ☐ g
☐ kg = 4000 g	☐ kg = 8000 g

Exercise 3

Practise your writing.   gram   kilogram

______________________   ______________________

______________________   ______________________

Activity

You will need: a 1 kg weight, balance scales, 6 objects

What to do:

1   Pick up the 1 kg weight. Hold it in your hand to feel how heavy it is.

2   Pick up each object your teacher has provided.

   Write down the name of the object and whether it *feels* heavier or lighter than 1 kg.

	Object	*Does it feel more or less than 1 kg?*	*Actual mass compared with 1 kg*
a			
b			
c			
d			
e			
f			

3   Place the 1 kg weight *gently* in one tray of the scale.

   One at a time, place each of your objects in the other tray.

   Record whether each object is actually heavier or lighter than 1 kg.

Discussion

Is the *size* of an object always a good indication of its *mass*?

The mass of a light object can be measured in grams using a digital scale.

The apple has mass 182 g.

Exercise 4

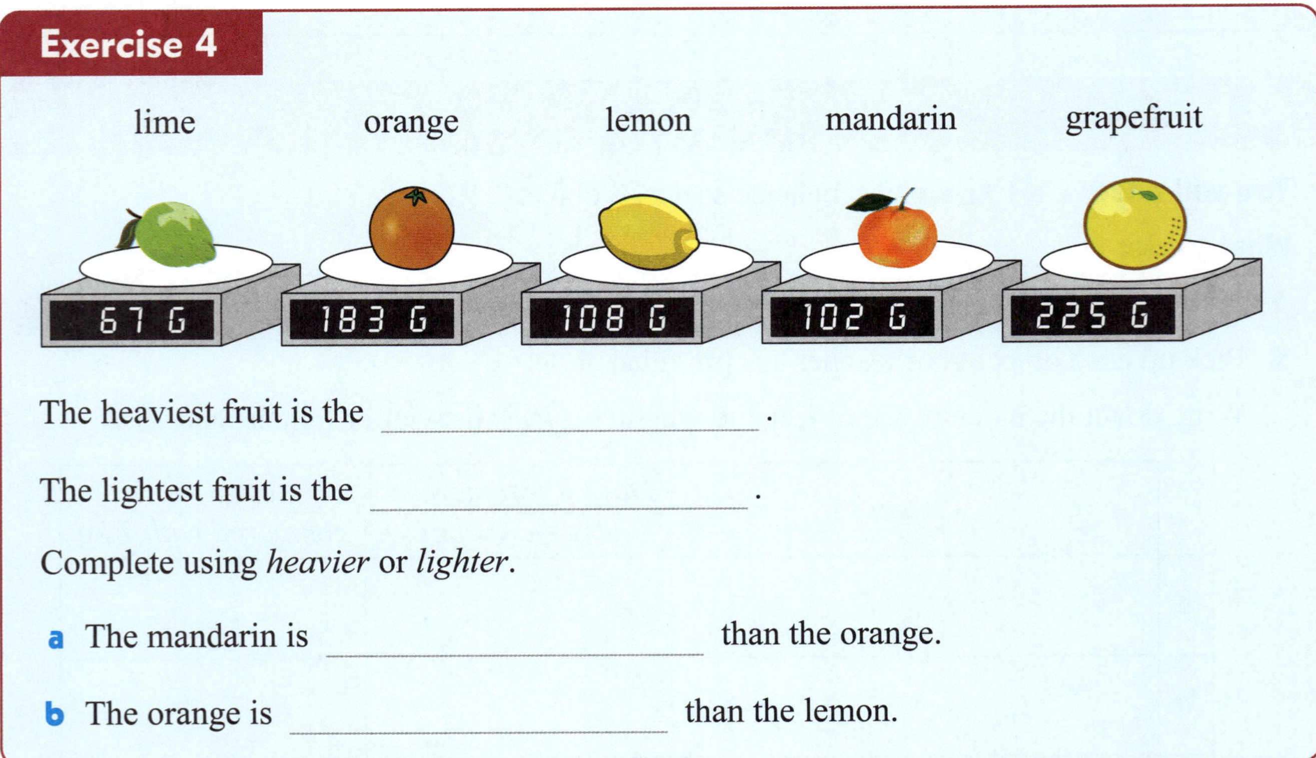

The heaviest fruit is the ________________________ .

The lightest fruit is the ________________________ .

Complete using *heavier* or *lighter*.

a The mandarin is ________________________ than the orange.

b The orange is ________________________ than the lemon.

Exercise 5

Write the items in order of mass.

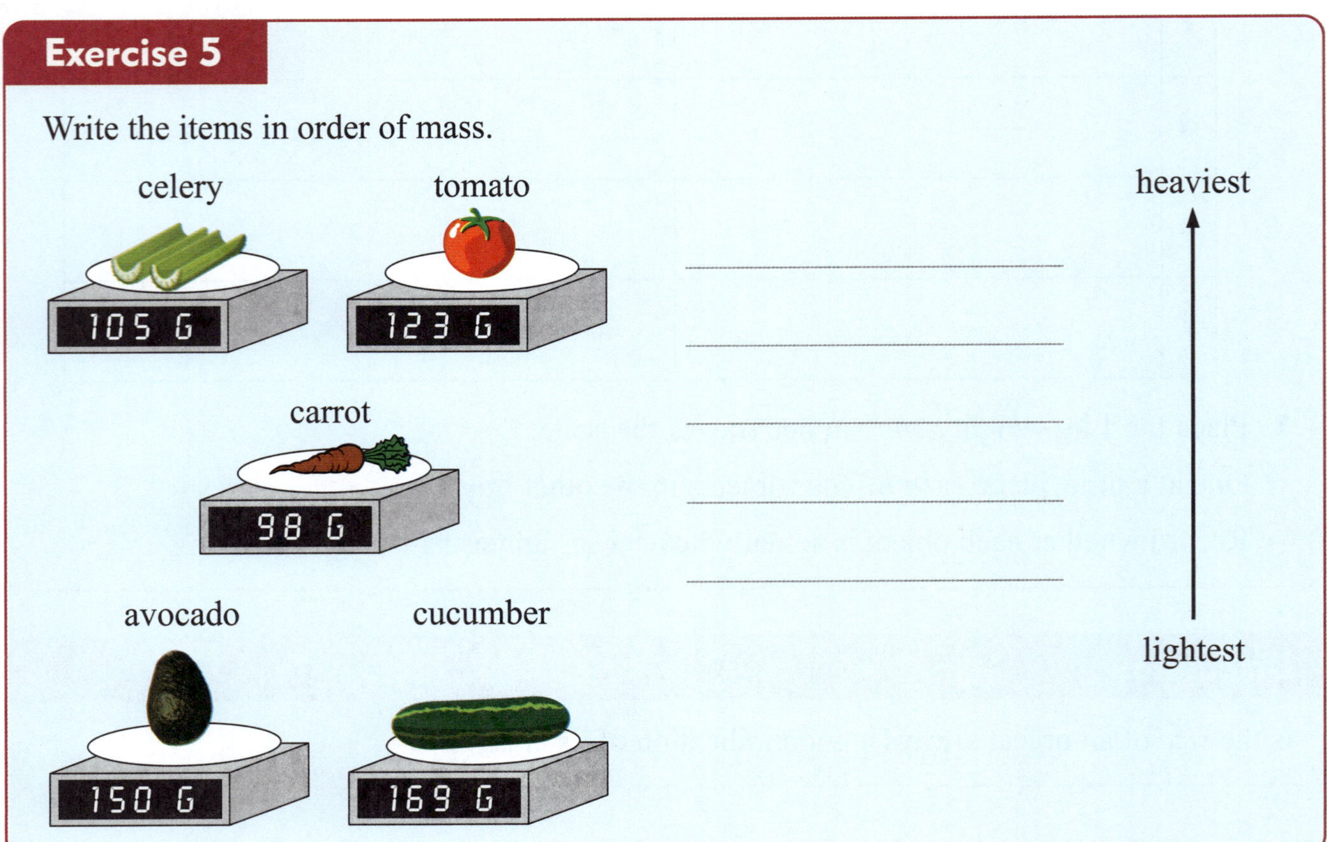

Game

Click on the icon to play a game about measuring mass!

Activity

You will need: digital scale, 5 small objects

What to do:

1 Use the digital scale to find the mass of each object.
Record your results in the table below.

	Object	Mass
a		
b		
c		
d		
e		

2 **a** The heaviest object is ________________________ .

b The lightest object is ________________________ .

Exercise 6

Joe has mass 37 kg. His sister Belinda has mass 44 kg.

a Who is heavier?   ____________________

b By how much is this person heavier?

____

c What is the total mass of Joe and Belinda?

____

Revision

1 Use *heavier* or *lighter* to complete each sentence:

a

The cube is

____

than the cone.

b

The pyramid is

____

than the cylinder.

2 Use *heavier* or *lighter* to complete each sentence:

a A house is ___________________ than a mountain.

b An eagle is ___________________ than a butterfly.

3 Complete:

$1 \text{ kg} = \boxed{} \text{ g}$

$4 \text{ kg} = \boxed{} \text{ g}$

$\boxed{} \text{ kg} = 9000 \text{ g}$

4

hamster   mouse   rat   gerbil

The heaviest animal is the ___________________ .

The lightest animal is the ___________________ .

The gerbil is ___________________ than the hamster.

The hamster is ___________________ than the mouse.

5 My cat has mass 4 kg. My dog has mass 18 kg.

My cat is _______ kg ___________________ than my dog.

Mark has 6 chickens.

Each chicken lays 2 eggs.

In total, Mark has

$$2 + 2 + 2 + 2 + 2 + 2 = 12 \text{ eggs.}$$

We can also write a **multiplication**:

$$6 \times 2 = 12$$

"Six times two equals twelve."

"Six multiplied by two is twelve."

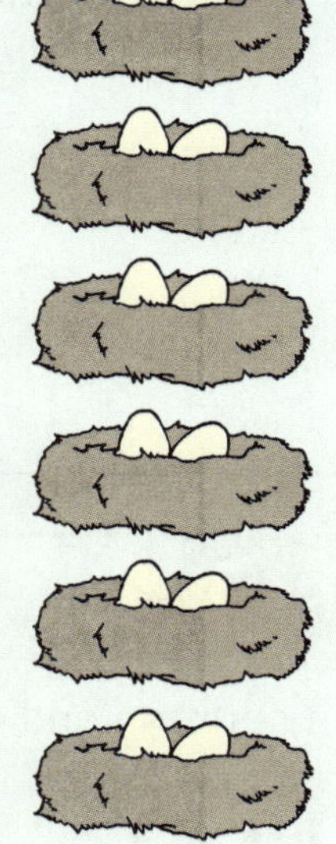

Exercise 1

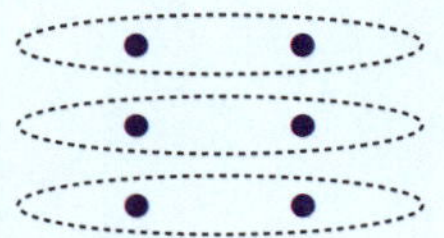

This array shows
$$2 + 2 + 2 = 6$$
$$3 \times 2 = 6$$

Write an addition and a multiplication for each array:

a

______ + ______ + ______ = ______

______ × ______ = ______

b

c 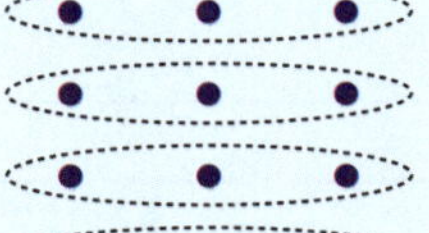

d 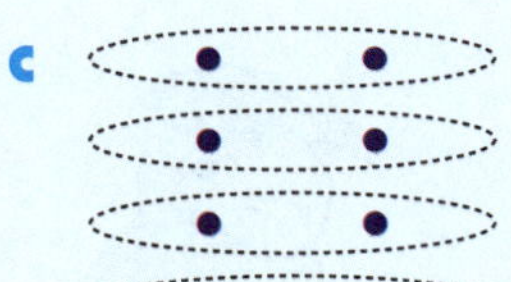

Exercise 2

$4 + 4 + 4 + 4 + 4 + 4$ can be written as 6×4.

6 lots of 4

Write as a multiplication:

a $3 + 3 =$ _____ $\times$ _____

b $5 + 5 + 5 + 5 =$ _____ $\times$ _____

c $7 + 7 + 7 =$ _____ $\times$ _____

d $6 + 6 + 6 + 6 + 6 =$ _____ $\times$ _____

e $4 + 4 + 4 + 4 =$ _____ $\times$ _____

Exercise 3

Write an addition and a multiplication for each array:

a 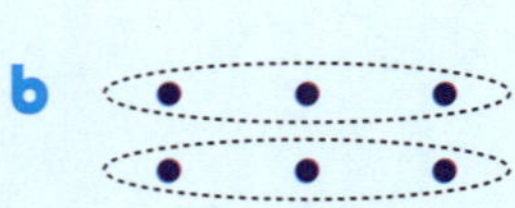

____

____

b 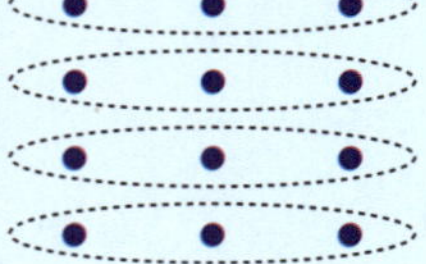

____

____

c

____

____

Exercise 4

Write as an addition and find the answer:

a $2 \times 7 =$ _____ $+$ _____ $=$ _____

b $3 \times 8 =$ _____ $+$ _____ $+$ _____ $=$ _____

c $4 \times 9 =$ _____ $+$ _____ $+$ _____ $+$ _____ $=$ _____

d $3 \times 10 =$ _____ $+$ _____ $+$ _____ $=$ _____

Exercise 5

Use the arrays to help write your **times tables**.

a 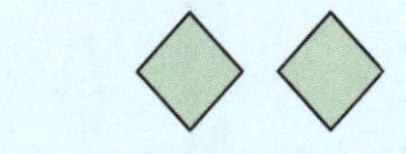

$1 \times 2 =$ ______

$2 \times 2 =$ ______

$3 \times 2 =$ ______

$4 \times 2 =$ ______

$5 \times 2 =$ ______

$6 \times 2 =$ ______

$7 \times 2 =$ ______

$8 \times 2 =$ ______

$9 \times 2 =$ ______

$10 \times 2 =$ ______

b

$1 \times 3 =$ ______

$2 \times 3 =$ ______

$3 \times 3 =$ ______

$4 \times 3 =$ ______

$5 \times 3 =$ ______

$6 \times 3 =$ ______

$7 \times 3 =$ ______

$8 \times 3 =$ ______

$9 \times 3 =$ ______

$10 \times 3 =$ ______

c 

$1 \times 4 =$ ______

$2 \times 4 =$ ______

$3 \times 4 =$ ______

$4 \times 4 =$ ______

$5 \times 4 =$ ______

$6 \times 4 =$ ______

$7 \times 4 =$ ______

$8 \times 4 =$ ______

$9 \times 4 =$ ______

$10 \times 4 =$ ______

d 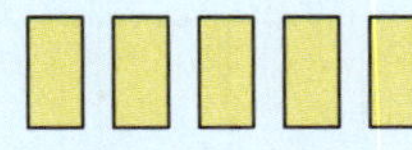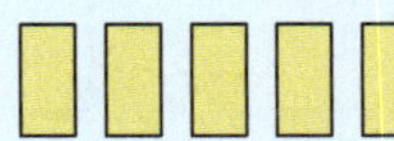

$1 \times 5 =$ ______

$2 \times 5 =$ ______

$3 \times 5 =$ ______

$4 \times 5 =$ ______

$5 \times 5 =$ ______

$6 \times 5 =$ ______

$7 \times 5 =$ ______

$8 \times 5 =$ ______

$9 \times 5 =$ ______

$10 \times 5 =$ ______

Exercise 6

There are _______ dots in total.

I can collect the dots in groups of 2.

There are _______ groups.

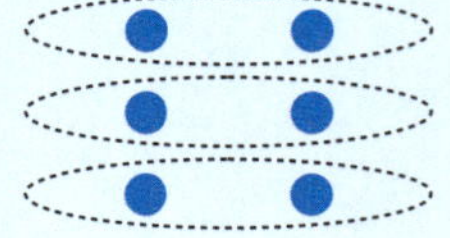

_______ × 2 = _______

I can collect the dots in groups of 3.

There are _______ groups.

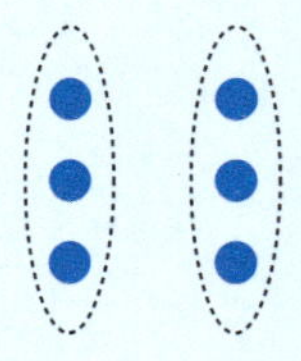

_______ × 3 = _______

Exercise 7

Use the dots to write *two* multiplications.

a.

☐ × ☐ = ☐

☐ × ☐ = ☐

b.

☐ × ☐ = ☐

☐ × ☐ = ☐

c.

☐ × ☐ = ☐

☐ × ☐ = ☐

d.

☐ × ☐ = ☐

☐ × ☐ = ☐

e.

☐ × ☐ = ☐

☐ × ☐ = ☐

f.

☐ × ☐ = ☐

☐ × ☐ = ☐

Discussion

What can we learn about multiplication from **Exercise 7**?

____

Exercise 8

Write matching multiplications.

$$5 \times 2 = 10$$
$$2 \times 5 = 10$$

a
$$3 \times 2 = 6$$
$$\boxed{} \times \boxed{} = \boxed{}$$

b
$$4 \times 3 = 12$$
$$\boxed{} \times \boxed{} = \boxed{}$$

c
$$3 \times 5 = 15$$
$$\boxed{} \times \boxed{} = \boxed{}$$

d
$$7 \times 2 = \boxed{}$$
$$\boxed{} \times \boxed{} = \boxed{}$$

e
$$8 \times 4 = \boxed{}$$
$$\boxed{} \times \boxed{} = \boxed{}$$

f
$$6 \times 5 = \boxed{}$$
$$\boxed{} \times \boxed{} = \boxed{}$$

g
$$9 \times 3 = \boxed{}$$
$$\boxed{} \times \boxed{} = \boxed{}$$

h
$$7 \times 5 = \boxed{}$$
$$\boxed{} \times \boxed{} = \boxed{}$$

Listening Activity

Listen carefully to the questions and write your answers in the spaces below.

1 ___________   **2** ___________   **3** ___________

4 ___________   **5** ___________   **6** ___________

7 ___________   **8** ___________   **9** ___________

10 ___________

Exercise 9

In our place value number system, the number 80 means 8 tens or 8×10.

$$8 \times 10 = 80$$

Complete your 10 times table.

$1 \times 10 = \underline{\hspace{2cm}}$ $6 \times 10 = \underline{\hspace{2cm}}$

$2 \times 10 = \underline{\hspace{2cm}}$ $7 \times 10 = \underline{\hspace{2cm}}$

$3 \times 10 = \underline{\hspace{2cm}}$ $8 \times 10 = \underline{\hspace{2cm}}$

$4 \times 10 = \underline{\hspace{2cm}}$ $9 \times 10 = \underline{\hspace{2cm}}$

$5 \times 10 = \underline{\hspace{2cm}}$ $10 \times 10 = \underline{\hspace{2cm}}$

Discussion

Can we use place values to complete these multiplications?

a $20 = 2 \times \underline{\hspace{2cm}}$ **b** $50 = 5 \times \underline{\hspace{2cm}}$

     $200 = 2 \times \underline{\hspace{2cm}}$ $500 = 5 \times \underline{\hspace{2cm}}$

     $2000 = 2 \times \underline{\hspace{2cm}}$ $5000 = 5 \times \underline{\hspace{2cm}}$

The results of the times tables are called **multiples**.

Exercise 10

The multiples of 2 are:

2, 4, 6, ______, ______, ______, ______, ______, ______, ______, ...

The multiples of 3 are:

3, 6, 9, ______, ______, ______, ______, ______, ______, ______, ...

The multiples of 5 are:

5, 10, ______, ______, ______, ______, ______, ______, ______, ______, ...

The multiples of 10 are:

10, 20, ______, ______, ______, ______, ______, ______, ______, ______, ...

Exercise 11

By counting up in 4s, we find the multiples of 4. They are shown in pink.

1	2	3	4	5	6	7	8	9	10
11	12	13	14	15	16	17	18	19	20
21	22	23	24	25	26	27	28	29	30
31	32	33	34	35	36	37	38	39	40

The multiples of 4 are:

4, 8, ______, ______, ______, ______, ______, ______, ______, ______, ...

Exercise 12

By counting up in 6s, find and colour the multiples of 6.

1	2	3	4	5	6	7	8	9	10
11	12	13	14	15	16	17	18	19	20
21	22	23	24	25	26	27	28	29	30
31	32	33	34	35	36	37	38	39	40
41	42	43	44	45	46	47	48	49	50
51	52	53	54	55	56	57	58	59	60

The multiples of 6 are:

6, ______, ______, ______, ______, ______, ______, ______, ______, ______, ...

The 6 times table is:

$1 \times 6 =$ ______ $6 \times 6 =$ ______

$2 \times 6 =$ ______ $7 \times 6 =$ ______

$3 \times 6 =$ ______ $8 \times 6 =$ ______

$4 \times 6 =$ ______ $9 \times 6 =$ ______

$5 \times 6 =$ ______ $10 \times 6 =$ ______

Exercise 13

By counting up in 7s, find and colour the multiples of 7.

1	2	3	4	5	6	7	8	9	10
11	12	13	14	15	16	17	18	19	20
21	22	23	24	25	26	27	28	29	30
31	32	33	34	35	36	37	38	39	40
41	42	43	44	45	46	47	48	49	50
51	52	53	54	55	56	57	58	59	60
61	62	63	64	65	66	67	68	69	70

The multiples of 7 are:

7, _______, _______, _______, _______, _______, _______, _______, _______, _______, ...

The 7 times table is:

$1 \times 7 =$ _______ $6 \times 7 =$ _______

$2 \times 7 =$ _______ $7 \times 7 =$ _______

$3 \times 7 =$ _______ $8 \times 7 =$ _______

$4 \times 7 =$ _______ $9 \times 7 =$ _______

$5 \times 7 =$ _______ $10 \times 7 =$ _______

Exercise 14

Complete these multiplication flowers.

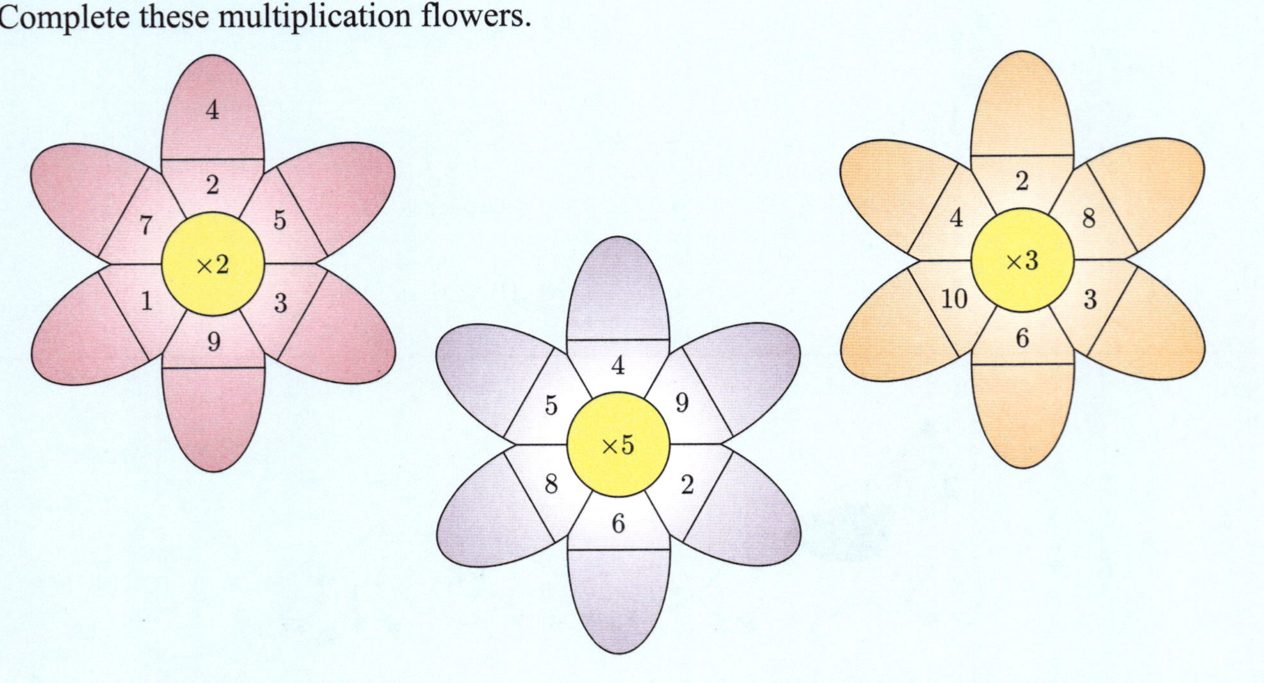

Exercise 15

By counting up on the number chart, complete the 8 and 9 times tables.

1	2	3	4	5	6	7	8	9	10
11	12	13	14	15	16	17	18	19	20
21	22	23	24	25	26	27	28	29	30
31	32	33	34	35	36	37	38	39	40
41	42	43	44	45	46	47	48	49	50
51	52	53	54	55	56	57	58	59	60
61	62	63	64	65	66	67	68	69	70
71	72	73	74	75	76	77	78	79	80
81	82	83	84	85	86	87	88	89	90
91	92	93	94	95	96	97	98	99	100

$1 \times 8 = $ _____ $6 \times 8 = $ _____

$2 \times 8 = $ _____ $7 \times 8 = $ _____

$3 \times 8 = $ _____ $8 \times 8 = $ _____

$4 \times 8 = $ _____ $9 \times 8 = $ _____

$5 \times 8 = $ _____ $10 \times 8 = $ _____

$1 \times 9 = $ _____ $6 \times 9 = $ _____

$2 \times 9 = $ _____ $7 \times 9 = $ _____

$3 \times 9 = $ _____ $8 \times 9 = $ _____

$4 \times 9 = $ _____ $9 \times 9 = $ _____

$5 \times 9 = $ _____ $10 \times 9 = $ _____

We can read 2×3 as:

- "2 times 3"
- "2 multiplied by 3"
- "the product of 2 and 3"

Exercise 16

Answer the multiplications.

a 2 times 2 is _____ .

b 3 multiplied by 5 is _____ .

c The product of 5 and 2 is _____ .

d 4 multiplied by 3 is _____ .

e The product of 7 and 2 is _____ .

f 9 times 3 equals _____ .

g 8 times 4 equals _____ .

h The product of 6 and 5 is _____ .

This **multiplication table** shows all of the times tables you have discovered.

×	1	2	3	4	5	6	7	8	9	10
1	1	2	3	4	5	6	7	8	9	10
2	2	4	6	8	10	12	14	16	18	20
3	3	6	9	12	15	18	21	24	27	30
4	4	8	12	16	20	24	28	32	36	40
5	5	10	15	20	25	30	35	40	45	50
6	6	12	18	24	30	36	42	48	54	60
7	7	14	21	28	35	42	49	56	63	70
8	8	16	24	32	40	48	56	64	72	80
9	9	18	27	36	45	54	63	72	81	90
10	10	20	30	40	50	60	70	80	90	100

The shaded lines show that $7 \times 4 = 28$.

You need to learn your times tables by heart.

Click the icon alongside for help with your times tables.

Exercise 17

Complete these multiplications.

a $3 \times 7 =$ _______

b $2 \times 8 =$ _______

c $10 \times 9 =$ _______

d $4 \times 6 =$ _______

e $7 \times 7 =$ _______

f $9 \times 8 =$ _______

g $8 \times 7 =$ _______

h $5 \times 9 =$ _______

i $6 \times 8 =$ _______

Exercise 18

Use the multiplication table to find multiplications which have the answer:

a 6

b 10

c 15

d 12

e 18

f 30

g 40

Exercise 19

Use multiplication to solve these questions.

a Lucy has 3 bowls. She places 5 strawberries into each bowl.

How many strawberries are there altogether?

☐ × ☐ = ☐

There are ☐ strawberries altogether.

b

Jonathon fixes 8 studs to the bottom of each rugby boot.

How many studs are there altogether?

☐ × ☐ = ☐

There are ☐ studs altogether.

c Hannah eats 3 pieces of fruit every day.

How many pieces of fruit does she eat in 9 days?

☐ × ☐ = ☐

Hannah eats ☐ pieces of fruit in 9 days.

Melanie has 3 empty boxes. They each contain 0 light globes.

Melanie has $3 \times 0 = 0$ light globes.

Bernard's store should sell boxes with 2 light globes in each, but he has sold out.

Bernard has $0 \times 2 = 0$ light globes.

Exercise 20

Complete each multiplication:

a $8 \times 0 =$ _______   **b** $0 \times 9 =$ _______   **c** $4 \times$ _______ $= 0$

Exercise 21

Complete the multiplication cobwebs.

a 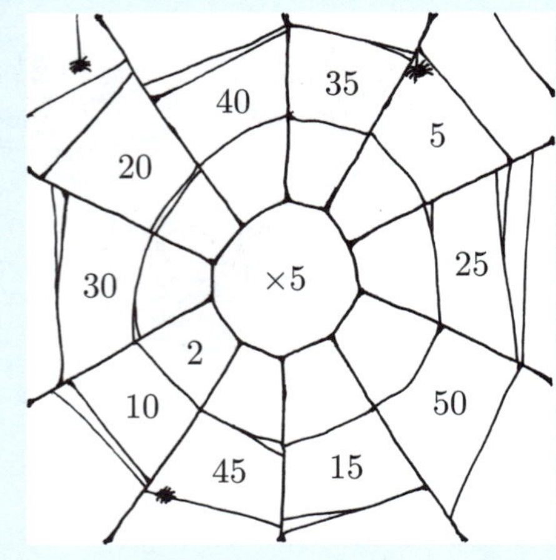

b 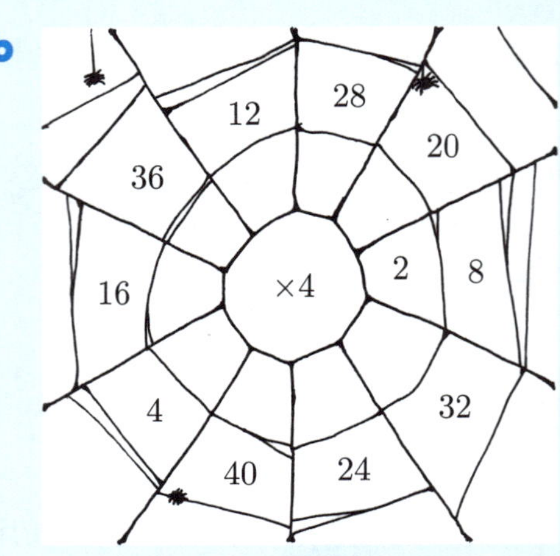

Exercise 22

Use multiplication to solve these questions.

a Andy swims 8 laps of a swimming pool each day.

In 4 days, Andy will swim ☐ × ☐ = ☐ laps.

In 6 days, Andy will swim ☐ × ☐ = ☐ laps.

In 10 days, Andy will swim ☐ × ☐ = ☐ laps.

b Mai reads 9 pages of her book each night.

In 2 days, Mai will read ☐ × ☐ = ☐ pages.

In 5 days, Mai will read ☐ × ☐ = ☐ pages.

In 7 days, Mai will read ☐ × ☐ = ☐ pages.

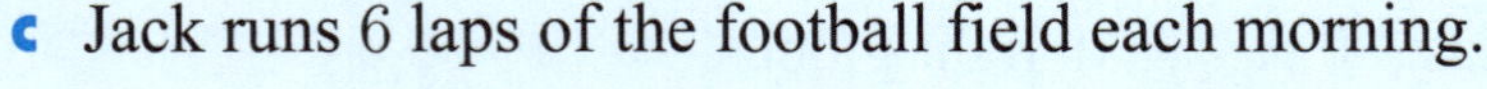

c Jack runs 6 laps of the football field each morning.

In 2 days, Jack will run ☐ × ☐ = ☐ laps.

In 4 days, Jack will run ☐ × ☐ = ☐ laps.

In 5 days, Jack will run ☐ × ☐ = ☐ laps.

In 9 days, Jack will run ☐ × ☐ = ☐ laps.

We know that 10 units make 1 ten

and 10 tens make 1 hundred.

When we multiply by 10, each digit moves one place to the *left*.

We write a zero to fill the empty units place.

$$14 \times 10 = 140$$

Exercise 23

Complete each multiplication:

a $8 \times 10 =$ _______

b $10 \times 10 =$ _______

c $30 \times 10 =$ _______

d $12 \times 10 =$ _______

e $70 \times 10 =$ _______

f $35 \times 10 =$ _______

g $43 \times 10 =$ _______

h $64 \times 10 =$ _______

i $89 \times 10 =$ _______

Puzzle

1 Philippa has 5 cats.

Each cat has 4 kittens.

How many cats does Philippa now have in total? _______

2 Jaime has 2 barns.

Each barn has 3 horses.

Each horse has 4 legs.

How many horse legs are there in total? _______

Challenge

Complete each multiplication:

a $28 \times 10 =$ _______

$28 \times 100 =$ _______

$28 \times 1000 =$ _______

b $61 \times 10 =$ _______

$61 \times 100 =$ _______

$61 \times 1000 =$ _______

Revision

1 Write an addition and a multiplication for each array:

a 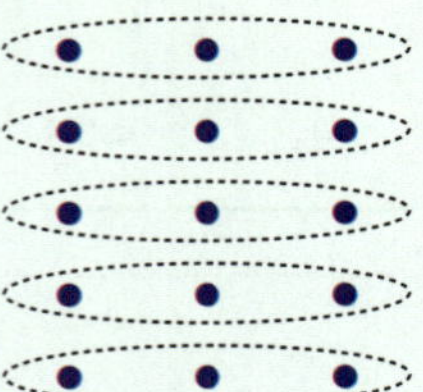

_____ + _____ = _____

_____ × _____ = _____

b 

____

____

2 Write as a multiplication and find the answer:

a $8 + 8 + 8 =$ _____ × _____ = _____

b $4 + 4 + 4 + 4 + 4 =$ _____ × _____ = _____

3 Write as an addition and find the answer:

a $5 \times 6 =$ _____ + _____ + _____ + _____ + _____ = _____

b $3 \times 9 =$ _____ + _____ + _____ = _____

4 Use the dots to write *two* multiplications.

a 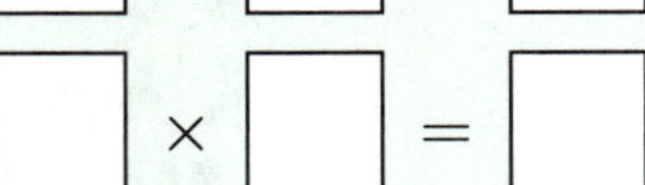 □ × □ = □

□ × □ = □

b  □ × □ = □

□ × □ = □

5 Complete the multiplications.

a $7 \times 3 =$ _____

b $2 \times 5 =$ _____

c $7 \times 2 =$ _____

d $6 \times 10 =$ _____

e $4 \times 8 =$ _____

f $9 \times 6 =$ _____

g $5 \times 5 =$ _____

h $8 \times 7 =$ _____

i $6 \times 4 =$ _____

j $10 \times 3 =$ _____

k $7 \times 9 =$ _____

l $8 \times 8 =$ _____

m $4 \times 7 =$ _____

n $6 \times 8 =$ _____

o $7 \times 7 =$ _____

p $5 \times 8 =$ _____

q $3 \times 9 =$ _____

r $10 \times 7 =$ _____

6 Write matching multiplications.

a
$$5 \times 6 = \boxed{}$$
$$6 \times \boxed{} = \boxed{}$$

b
$$9 \times 4 = \boxed{}$$
$$\boxed{} \times \boxed{} = \boxed{}$$

7 Complete each multiplication:

a 2 times 8 is __________ .

b 3 multiplied by 2 is __________ .

c 6 times 6 equals __________ .

d The product of 3 and 5 is __________ .

e 8 multiplied by 6 is __________ .

f The product of 9 and 10 is __________ .

8 Complete these multiplication flowers.

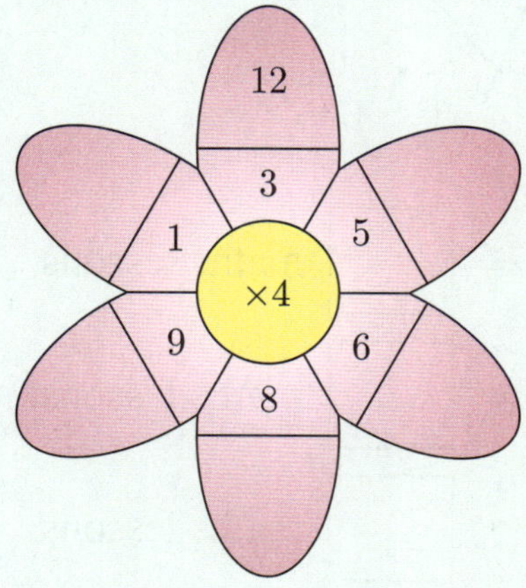

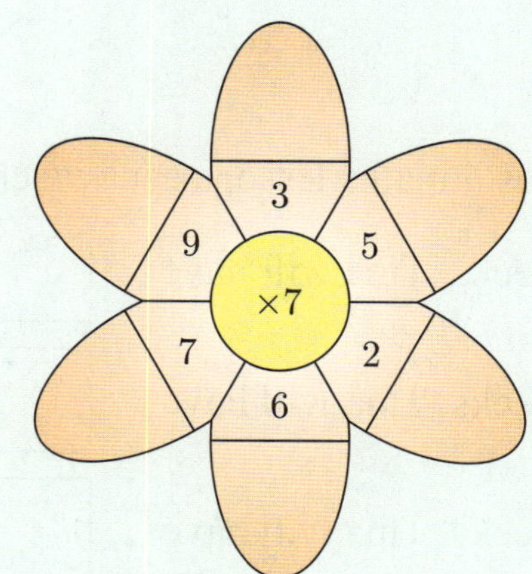

9 a The multiples of 8 are:

8, 16, ______ , ______ , ______ , ______ , ______ , ______ , ______ , ______ , ...

b The multiples of 9 are:

9, 18, ______ , ______ , ______ , ______ , ______ , ______ , ______ , ______ , ...

10 Write down multiplications which have the answer:

a 16 ________________ ________________ ________________

b 20 ________________ ________________ ________________

c 24 ________________ ________________ ________________

11 Complete each multiplication:

a $7 \times 0 =$ ______   **b** $0 \times 5 =$ ______   **c** ______ $\times 6 = 0$

12 George practises the drums 4 days each week.

How many days does George practise in a 10 week term?

$\boxed{} \times \boxed{} = \boxed{}$

George practises $\boxed{}$ days in a 10 week term.

13 Complete the multiplication cobweb.

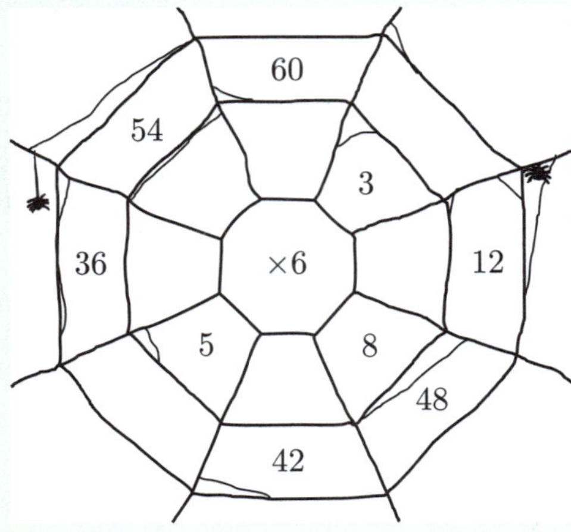

14 Tina has 5 maths lessons each week.

In 3 weeks, Tina will have $\boxed{} \times \boxed{} = \boxed{}$ maths lessons.

In 5 weeks, Tina will have $\boxed{} \times \boxed{} = \boxed{}$ maths lessons.

In 8 weeks, Tina will have $\boxed{} \times \boxed{} = \boxed{}$ maths lessons.

In 10 weeks, Tina will have $\boxed{} \times \boxed{} = \boxed{}$ maths lessons.

15 Complete each multiplication:

a $18 \times 10 = $ _______   **b** $40 \times 10 = $ _______   **c** $37 \times 10 = $ _______

CHAPTER 10: DATA HANDLING

When we look at the world around us, we often **count** things.

The information we collect is called **data**.

Exercise 1

Belinda designed a new shield for her favourite superhero.

The design has:

[ ] crowns

[ ] circles

[ ] stars

Mrs Gupta counted the cutlery in her kitchen.

She recorded the data using a **tally**.

She drew one **tally mark** for each item.

| represents 1 item.

⦀⦀ represents 5 items.

Cutlery	Tally
Knives	⦀⦀ |
Forks	⦀⦀ ||
Spoons	⦀⦀ ⦀⦀ |

Mrs Gupta counted 6 knives, 7 forks, and 11 spoons.

In total, Mrs Gupta counted $6 + 7 + 11 = 24$ pieces of cutlery.

Discussion

Why do you think it is helpful for Mrs Gupta to use a tally?

Exercise 2

Mr Simpson counted the tools in his workshop.

He recorded the data using a tally.

Tool	Tally
Spanners	⫲⫲⫲⫲ ‖‖‖‖
Screwdrivers	⫲⫲⫲⫲ ⫲⫲⫲⫲ ‖‖‖
Chisels	‖‖‖
Paintbrushes	⫲⫲⫲⫲ ‖

Mr Simpson counted:

[   ] spanners

[   ] screwdrivers

[   ] chisels

[   ] paintbrushes

Mr Simpson counted [   ] tools in total.

Exercise 3

The favourite drinks of students in Lucy's class are shown below using

 = milk,    = orange juice,   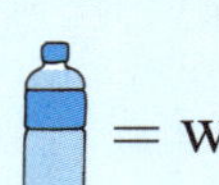 = water.

a Use a tally to record the results.

Drink	Tally
Milk	
Orange juice	
Water	

b Which drink is the most popular? __________________________________

c How many students selected water? __________________

d In total, how many students were asked their favourite drink? __________________

e How many *more* students selected orange juice than selected milk?

____

Exercise 4

Sam asked his classmates the question, "What is your favourite colour?"

These are the colours his classmates chose.

B	R	Y	B	O	R	G
B	G	G	R	G	O	R
R	R	B	R	R	G	R
R	O	O	G	R	G	P
R	Y	P				

B = blue
R = red
Y = yellow
O = orange
G = green
P = purple

a Use a tally to record the results.

b How many students chose blue?

____

c The most popular favourite colour was

_______________________ .

d How many students chose either yellow *or* orange?

____

Colour	Tally
Blue	

e Sam did not fill in the tally correctly. His tally is shown alongside.

Describe the mistakes that Sam made.

Colour	Tally
Blue	||||
Red	|||||||||||
Yellow	|
Orange	卌
Green	卌 |||
Pink	||

____

____

____

____

____

____

____

Activity

What to do:

1 In order to collect data, you need to ask a question.
For each of these questions, discuss what answers people might give. Write a list for each question.

a What is your hair colour?

Hair colour	Number
Total	

b How many siblings do you have?

Siblings	Number
Total	

c How did you travel to school today?

Method of travel	Number
Total	

2 Your teacher will prepare tables so that every student in the class can record their answers. Every student can place *one* tally mark for each question.

3 Record the number of students for each answer in the tables above.

4 Add the numbers for each question. Why should your totals all be the same?

5 The most common answers were:

a hair colour

b siblings

c method of travel to school

6 If you asked the same questions to another class, would you get the same answers?

I think this because

A **pictograph** uses pictures to show data.

The **key** tells us what each picture represents.

Mr McDonald counted the car colours in the staff car park.

He counted 5 white cars, 2 blue cars, 3 black cars, and 4 red cars.

He displayed his results in a pictograph.

Car colours in the staff car park

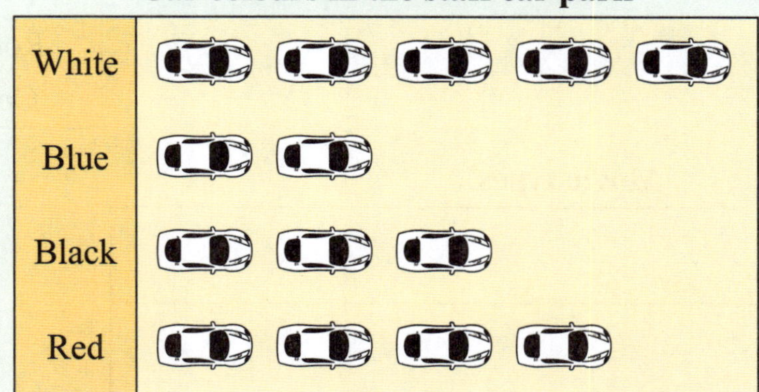

Key: = 1 car

The most common car colour was white.

Exercise 5

Jenny asked all the students in her class to choose their favourite big cat.

She drew her data on a pictograph, including her own choice.

Favourite big cat

Cheetah	🏃 🏃 🏃 🏃 🏃 🏃
Jaguar	🏃
Leopard	🏃 🏃
Lion	🏃 🏃 🏃 🏃
Tiger	🏃 🏃 🏃 🏃 🏃 🏃 🏃

Key: 🏃 = 1 student

a Which was the most popular choice? _______________

b How many students chose leopard? _______________

c How many more students chose cheetah than chose lion? _______________

d In total, how many students are in Jenny's class? _______________

Exercise 6

Dylon recorded the types of movies showing at a cinema.

Draw a pictograph to display Dylon's results.

Movie type	Tally
Action	|||
Science fiction	卌	
Romance	卌
Comedy	||

Movie types

Action	
Science fiction	
Romance	
Comedy	

Key: 🍿 = 1 movie

Exercise 7

Daniel drew this pictograph to show how many spiders he saw in the garden each day.

He used the key 🕷 = 2 spiders.

Spiders in the garden

Monday	🕷 🕷 🕷
Tuesday	🕷 🕷 🕷 🕷 🕷
Wednesday	🕷 🕷 🕷
Thursday	🕷 🕷
Friday	🕷 🕷 🕷 🕷

Key: 🕷 = 2 spiders

a On Monday, Daniel saw

$2 + 2 + 2 =$ ______ spiders.

b How many spiders did Daniel see on Tuesday?

____

c On which day did Daniel see 4 spiders?

____

Discussion

If the key is 🕷 = 2 spiders,   how would you show:

- 1 spider _______________________
- 5 spiders? _______________________

Exercise 8

Kara has a flower store. She asked 30 people to choose their favourite flower. Her results are shown alongside.

Complete this pictograph to display Kara's results.

Flower	Number
Rose	8
Tulip	5
Daisy	10
Lily	4
Orchid	3

Favourite flower

Rose	👤 👤 👤 👤
Tulip	👤 👤 👤
Daisy	👤 👤 👤 👤 👤
Lily	
Orchid	

Key: 👤 = 2 people

Exercise 9

Paul counted the different belt colours of contestants in a karate tournament.

He displayed his data using a **bar graph**.

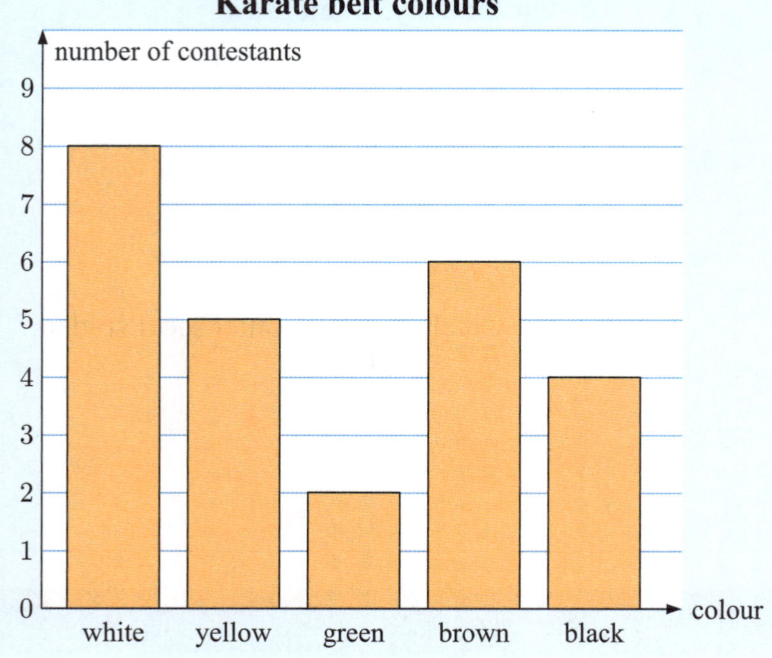

a Paul counted:

   ☐ white belts   ☐ yellow belts   ☐ green belts

   ☐ brown belts   ☐ black belts

b Were there more brown belts than yellow belts? _______________

   I know this because the brown belt bar is _______________ than the yellow belt bar.

c Were there fewer white belts than black belts? _______________

   I know this because the white belt bar is _______________ than the black belt bar.

Exercise 10

This bar graph shows the sandwiches bought in a café one lunchtime.

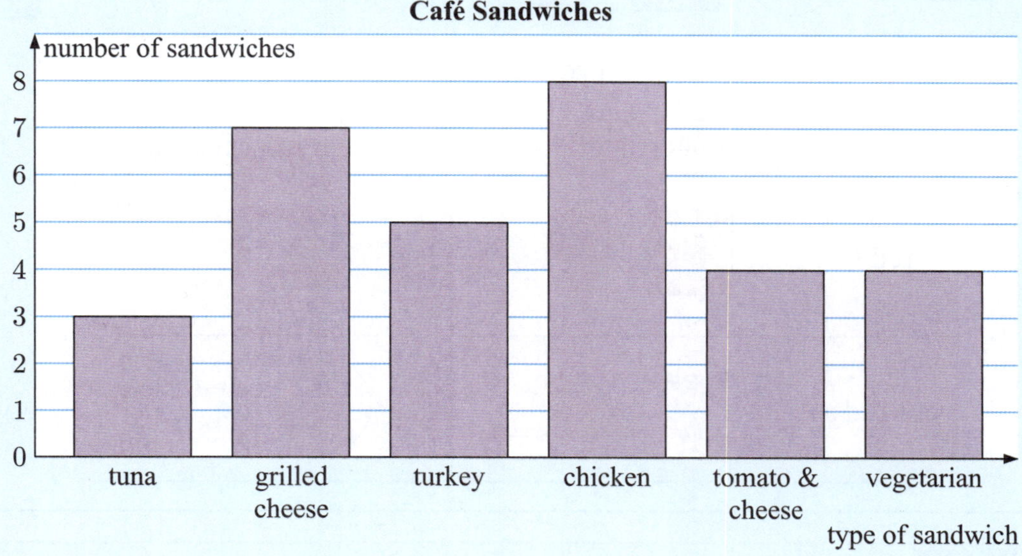

a The height of each bar shows the number of _______________________________

____ .

b The most popular sandwich was _______________________________________ .

c How many turkey sandwiches were bought? _______________________

d Were vegetarian sandwiches more popular than tuna sandwiches? _______________

e How many more chicken sandwiches were bought than vegetarian sandwiches?

____

f How many sandwiches with cheese were bought? _______________________

g How many sandwiches were bought in total? _______________________

Discussion

Compare a pictograph with a bar chart.

Which type of graph do you think is:

- easier to draw
- easier to understand
- nicer to look at?

Look at these shapes Ishaan has collected:

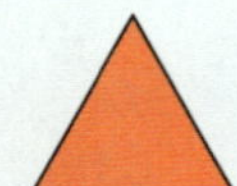 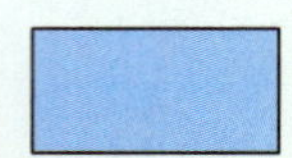 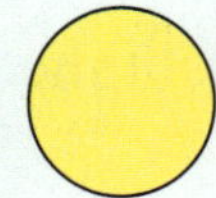 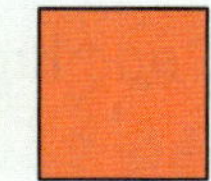 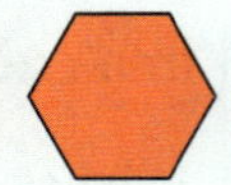 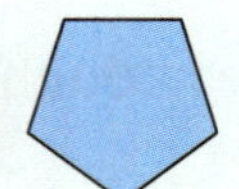 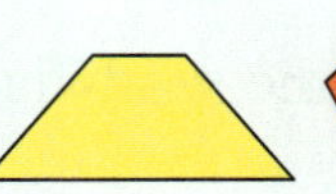  

Ishaan organised the shapes using a **Carroll diagram**:

	quadrilateral	not a quadrilateral
red	■ ◆	▲ ⬡
not red	▭ ▱	● ⬠

Exercise 11

Kit and Bella were asked if they like eating certain foods.

The results are shown in this Carroll diagram:

	Bella likes eating	Bella does not like eating
Kit likes eating	apple, cheese, strawberry	ice cream, bread
Kit does not like eating	banana, carrot, pizza	broccoli, egg

a Does Bella like eating cheese? ___________

b Does Kit like eating bananas? ___________

c Who likes eating ice cream? ___________

d Who likes eating carrots? ___________

e Which of these foods is liked by both Kit *and* Bella?

Circle the correct answer.   bread   apple   egg

f How many of the foods are not liked by either person? ___________

Activity

You will need: scissors, glue or sticky tape

What to do:

1 Click the icon, then print and cut out the objects.

2 Stick each object into the correct box of this Carroll diagram:

	has 4 legs	does not have 4 legs
living		
not living		

3 Draw one other object of your choice in each box of the diagram.

Look again at these shapes Ishaan has collected:

Ishaan has now organised the shapes using a **Venn diagram**:

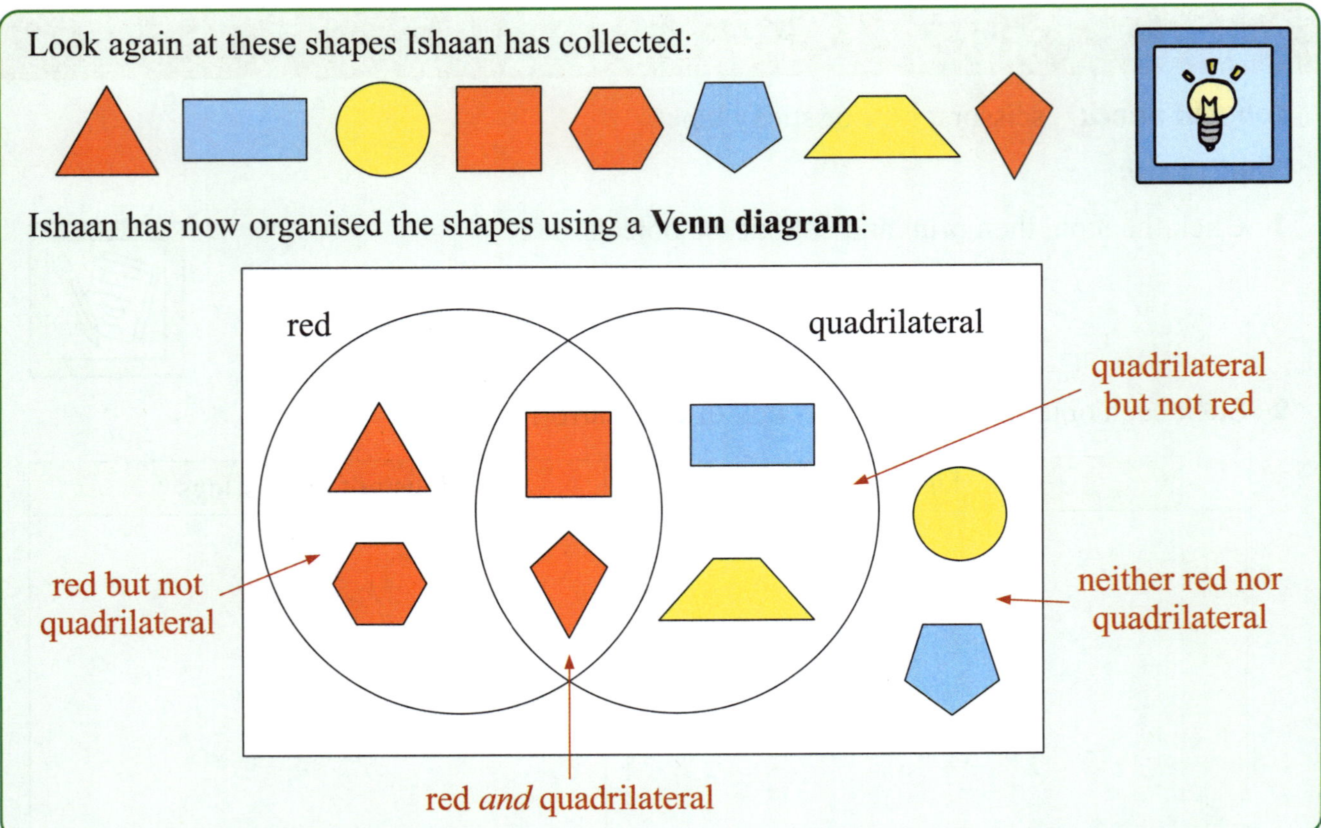

Exercise 12

Maddy asked her friends two questions:

- Do you like swimming?
- Do you like running?

She organised the results using a Venn diagram.

a Does Chin like running?

____

b Does Alex like swimming?   _______________

c How many friends like both swimming *and* running?

____

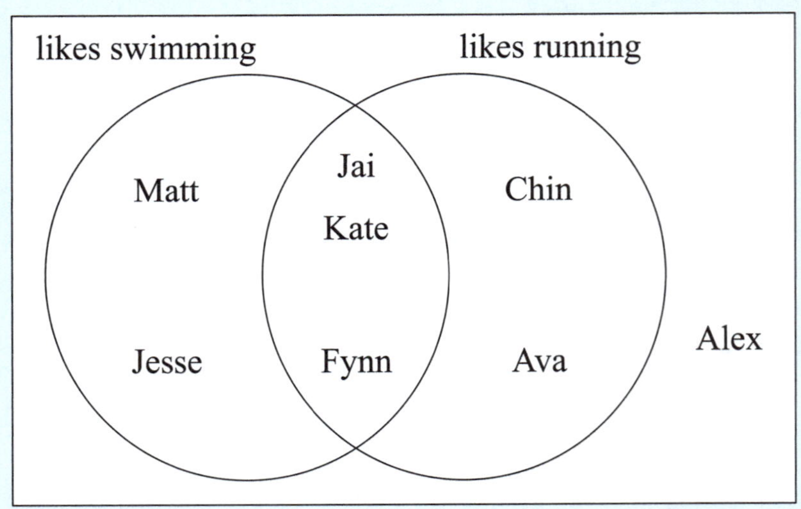

d Which two friends like swimming but not running?

_______________ and _______________

e Maddy herself likes running but does not like swimming. Add Maddy's name to the Venn diagram.

Exercise 13

Write all of the items on this shopping list in the Venn diagram.

peas   soap

rice   yoghurt

tissues   pasta

pens   shampoo

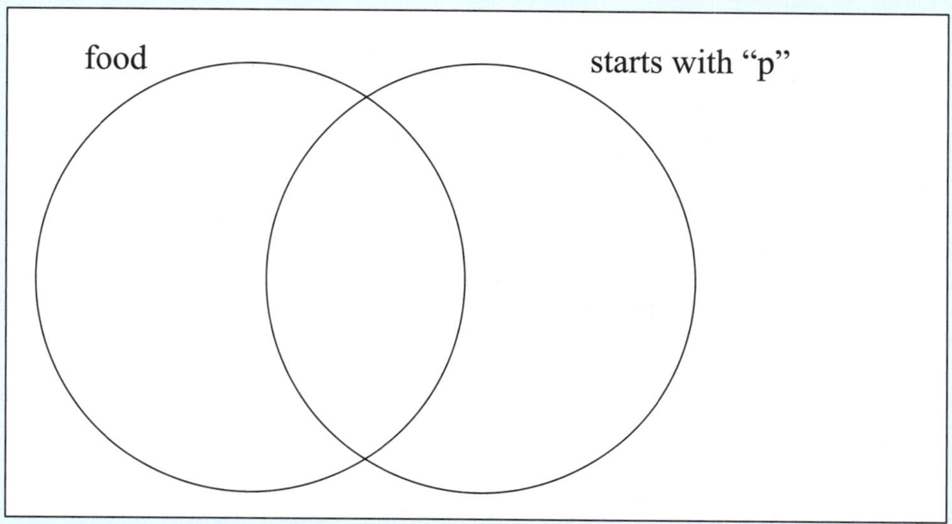

Revision

1 Martha counted the clothes in her wardrobe. She recorded the data as a tally.

Item	Tally
T-shirt	‖‖ ‖‖ |||
trousers	‖‖ ||||
skirt	|||
coat	‖‖	
dress	‖‖ ||

a How many T-shirts does Martha own? _______

b How many *more* trousers does Martha own than skirts?

____

c How many *fewer* coats does Martha own than trousers?

____

d Does Martha own more trousers or dresses? Write your answer in a sentence.

____

e In total, how many items does Martha have in her wardrobe? _______

2 **a** Record the colour of each block using a tally.

Colour	Tally
Blue	
Red	
Yellow	

b How many red blocks are there? _______________

c How many more blue blocks are there than yellow blocks?

____

d How many blocks are there in total? _______________

3 Olivia recorded the flavour of juice boxes sold at the school shop in one day.

Complete this pictograph to display the results.

Flavour	Tally
Apple	卌 l
Orange	|||
Pineapple	卌
Grape	||

Juice boxes bought at school shop

Apple	
Orange	
Pineapple	
Grape	

Key:  = 1 juice box

4 Jim drew this pictograph to show the results of his football team's matches.

Football results

Win	⚽ ⚽ ⚽
Draw	⚽ ◖
Loss	⚽ ⚽ ⚽ ◖

Key: ⚽ = 2 matches

a How many matches did the team win? _______________

b How many matches did the team lose? _______________

c The team lost _______ more matches than they drew.

5 This bar graph shows the colours of dogs in a park.

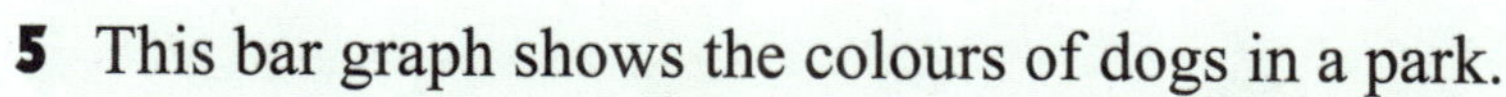

a The most common dog colour was ________________________ .

b There were _________ more yellow dogs than white dogs.

c There were _________ less black dogs than mixed colour dogs.

d There were _________ dogs in total in the park.

6 Aiden visited a petting zoo. He organised the animals he saw into a Carroll diagram:

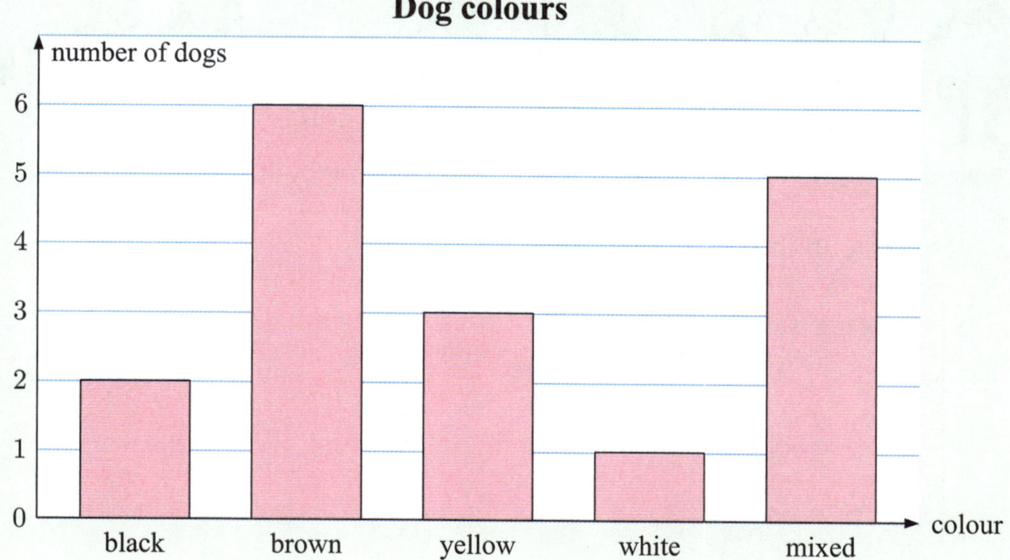

	fed	did not feed
petted		
did not pet		

a Did Aiden feed the duck? ___________

b Did Aiden pet the sheep? ___________

c Which animals did Aiden feed but not pet?

________________________ and ________________________

7

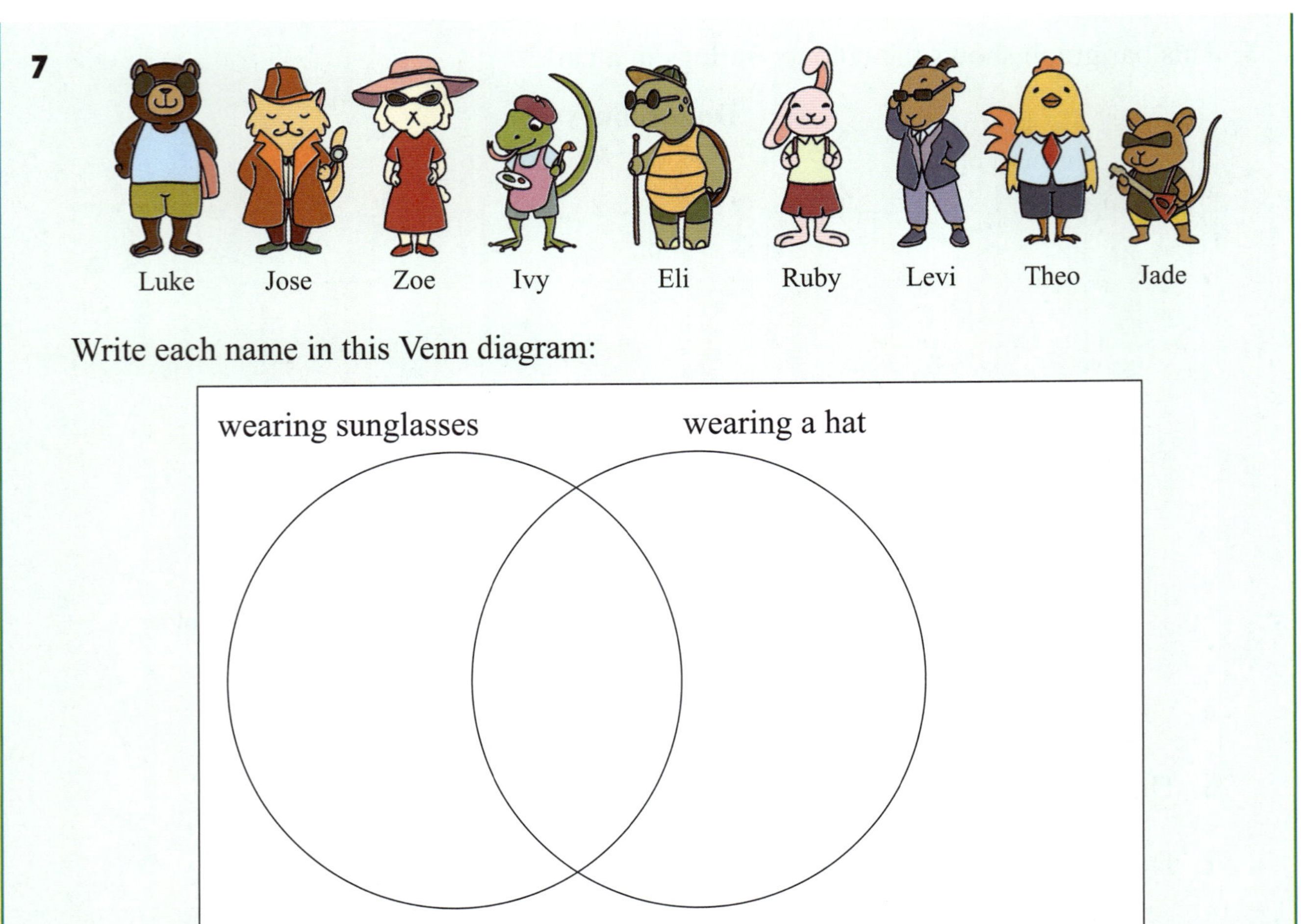

Write each name in this Venn diagram:

Hannah has 15 apples. She shares them **equally** with Lucy and Jane.

Hannah's box   Lucy's box   Jane's box

Each person now has **5** apples.

Hannah has **divided** the 15 apples into 3 equal groups of 5.

Exercise 1

Describe the division:

a _______ has been divided into

_______ equal groups of _______ .

b _______ has been divided into

_______ equal groups of _______ .

Exercise 2

Use counters to share the amounts, then draw your answer.

a Share 12 into 3 equal groups.

b Share 16 into 4 equal groups.

12 is shared into 3 equal groups of _______ .

16 is shared into 4 equal groups of _______ .

The symbol $\div$ means "divide".

8 has been divided into 2 equal groups of 4.

$$8 \div 2 = 4$$

"eight divided by two equals four"

Exercise 3

What division is shown by the diagram?

a

b

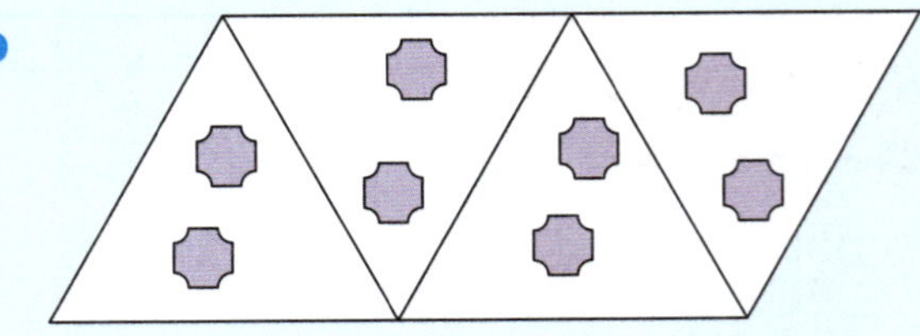

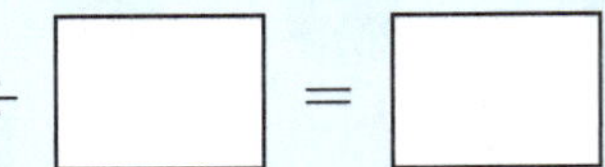

c

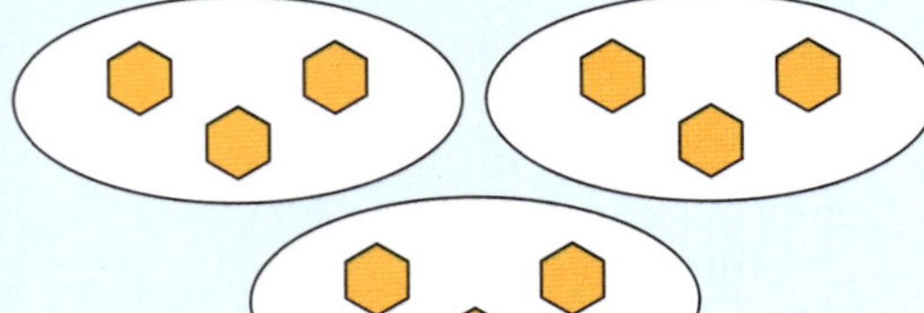

d

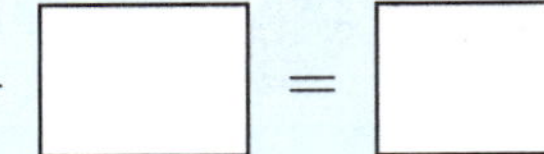

e

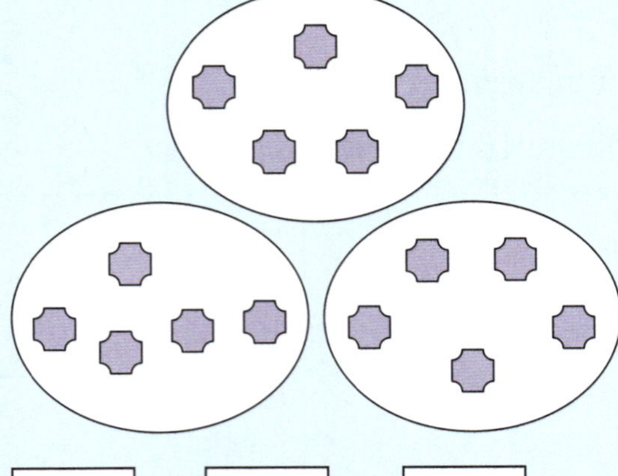

f

Exercise 4

Use a diagram to show the division and find the answer.

Make sure:

- each group contains the same number
- you have the correct total number.

a $4 \div 2 = $ _______

b $15 \div 3 = $ _______

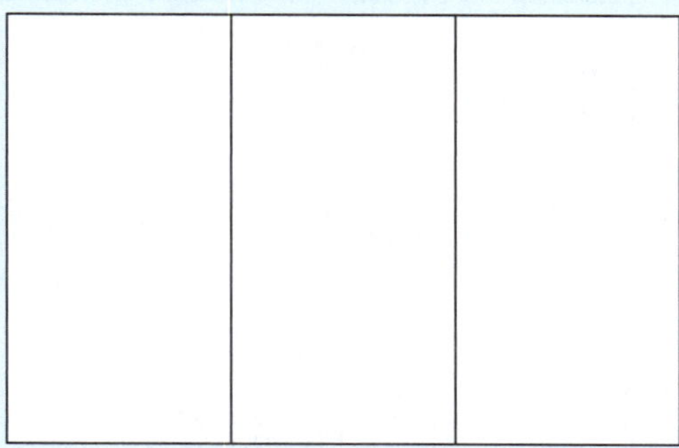

c $12 \div 4 = $ _______

d $20 \div 5 = $ _______

e $21 \div 3 = $ _______

f $24 \div 4 = $ _______

Division is the **opposite** of multiplication.

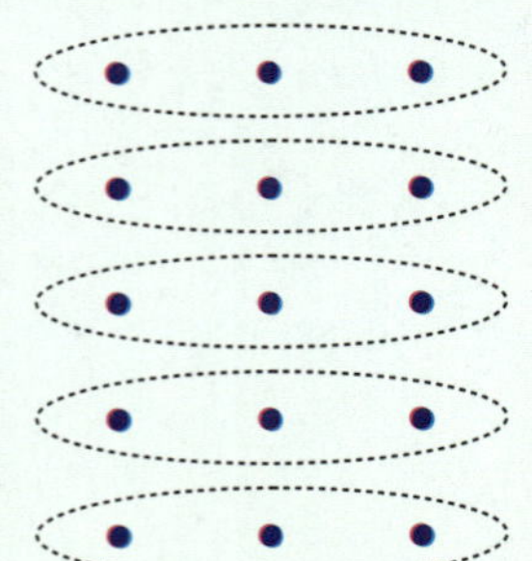

The array shows:

- 5 groups of 3 makes 15 in total.

  $5 \times 3 = 15$

- 15 is divided into 5 equal groups of 3.

  $15 \div 5 = 3$

Exercise 5

Use each array to write a multiplication and a division:

a 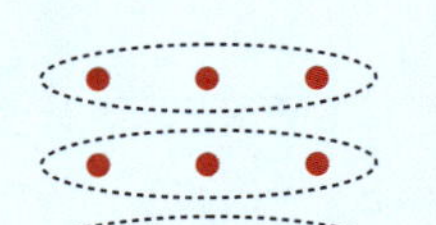

$\boxed{} \times \boxed{} = \boxed{}$

$\boxed{} \div \boxed{} = \boxed{}$

b 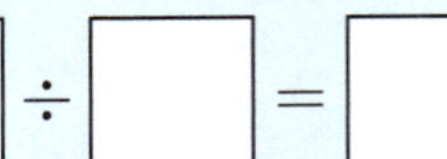

$\boxed{} \times \boxed{} = \boxed{}$

$\boxed{} \div \boxed{} = \boxed{}$

c 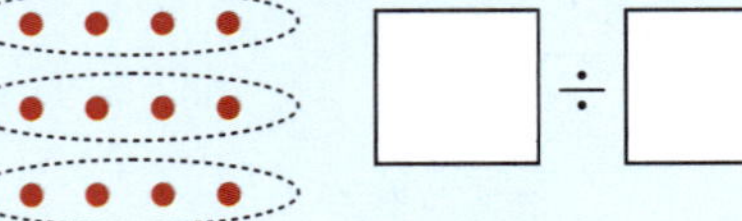

$\boxed{} \times \boxed{} = \boxed{}$

$\boxed{} \div \boxed{} = \boxed{}$

d 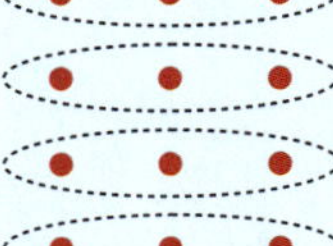

$\boxed{} \times \boxed{} = \boxed{}$

$\boxed{} \div \boxed{} = \boxed{}$

e 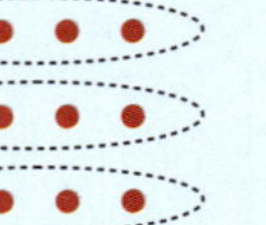

$\boxed{} \times \boxed{} = \boxed{}$

$\boxed{} \div \boxed{} = \boxed{}$

f 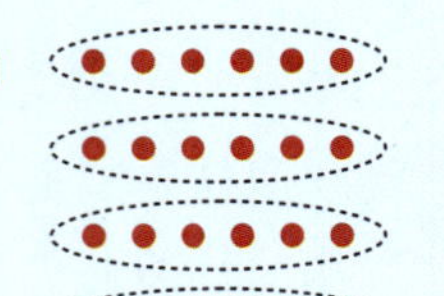

$\boxed{} \times \boxed{} = \boxed{}$

$\boxed{} \div \boxed{} = \boxed{}$

g 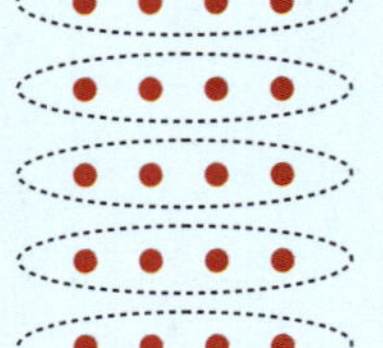

$\boxed{} \times \boxed{} = \boxed{}$

$\boxed{} \div \boxed{} = \boxed{}$

h 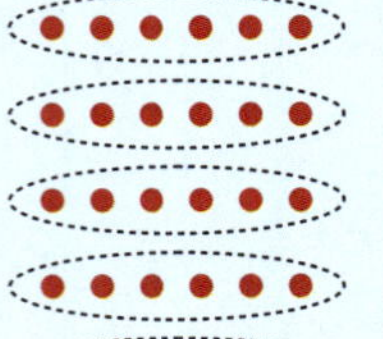

$\boxed{} \times \boxed{} = \boxed{}$

$\boxed{} \div \boxed{} = \boxed{}$

Discussion

How can your times tables help with division?

×	1	2	3	4	5	6	7	8	9	10
1	1	2	3	4	5	6	7	8	9	10
2	2	4	6	8	10	12	14	16	18	20
3	3	6	9	12	15	18	21	24	27	30
4	4	8	12	16	20	24	28	32	36	40
5	5	10	15	20	25	30	35	40	45	50
6	6	12	18	24	30	36	42	48	54	60
7	7	14	21	28	35	42	49	56	63	70
8	8	16	24	32	40	48	56	64	72	80
9	9	18	27	36	45	54	63	72	81	90
10	10	20	30	40	50	60	70	80	90	100

To find $32 \div 8$, look down the column for the 8 times table until you find 32.

$$8 \times 4 = 32$$
$$\text{so} \quad 32 \div 8 = 4$$

Exercise 6

Use the multiplication table to solve each division:

a $16 \div 8 = \underline{\hphantom{xxx}}$

b $9 \div 9 = \underline{\hphantom{xxx}}$

c $28 \div 4 = \underline{\hphantom{xxx}}$

d $18 \div 9 = \underline{\hphantom{xxx}}$

e $60 \div 10 = \underline{\hphantom{xxx}}$

f $15 \div 3 = \underline{\hphantom{xxx}}$

g $49 \div 7 = \underline{\hphantom{xxx}}$

h $27 \div 3 = \underline{\hphantom{xxx}}$

i $45 \div 5 = \underline{\hphantom{xxx}}$

j $63 \div 7 = \underline{\hphantom{xxx}}$

k $42 \div 6 = \underline{\hphantom{xxx}}$

l $72 \div 8 = \underline{\hphantom{xxx}}$

m $80 \div 8 = \underline{\hphantom{xxx}}$

n $64 \div 8 = \underline{\hphantom{xxx}}$

o $54 \div 9 = \underline{\hphantom{xxx}}$

p $48 \div 6 = \underline{\hphantom{xxx}}$

q $35 \div 7 = \underline{\hphantom{xxx}}$

r $90 \div 9 = \underline{\hphantom{xxx}}$

s $36 \div 4 = \underline{\hphantom{xxx}}$

t $56 \div 8 = \underline{\hphantom{xxx}}$

u $100 \div 10 = \underline{\hphantom{xxx}}$

We can use times tables to write related multiplications and divisions.

$$6 \times 8 = 48$$
$$8 \times 6 = 48$$
$$48 \div 6 = 8$$
$$48 \div 8 = 6$$

Exercise 7

Complete these related multiplications and divisions.

a
$3 \times 2 = $ _____
$2 \times 3 = $ _____
_____ $\div 3 = 2$
_____ $\div 2 = 3$

b
$4 \times 3 = $ _____
$3 \times 4 = $ _____
_____ $\div 4 = 3$
_____ $\div 3 = 4$

c
$10 \times 5 = $ _____
$5 \times 10 = $ _____
_____ $\div 10 = 5$
_____ $\div 5 = 10$

d
$6 \times 4 = $ _____
$4 \times 6 = $ _____
_____ $\div 6 = $ _____
_____ $\div 4 = $ _____

e
$2 \times$ _____ $= 16$
_____ $\times 2 = 16$
$16 \div 2 = $ _____
$16 \div$ _____ $= 2$

f
$9 \times$ _____ $= 27$
_____ $\times 9 = 27$
$27 \div 9 = $ _____
$27 \div$ _____ $= 9$

g
$7 \times$ _____ $= 35$
_____ $\times 7 = 35$
$35 \div 7 = $ _____
$35 \div$ _____ $= 7$

h
_____ $\times 6 = 54$
$6 \times$ _____ $= 54$
$54 \div$ _____ $= 6$
$54 \div 6 = $ _____

i
_____ $\times 8 = 72$
$8 \times$ _____ $= 72$
$72 \div$ _____ $= 8$
$72 \div 8 = $ _____

Discussion

What is the result if we divide a number by itself?   _____

What is the result if we divide a number by 1?   _____

Exercise 8

Use division to solve these questions.

a Ena has 32 mangoes. She shares them equally between 4 friends.

How many mangoes does each person receive?

Each person receives [ ] mangoes.

b In 8 packets there are a total of 48 cookies.

How many cookies are in each packet?

[ ] ÷ [ ] = [ ]

There are [ ] cookies in each packet.

Puzzle

Use these numbers only to write as many multiplications and divisions as you can.

2	3	4	6	12

Revision

1 What division is shown by the diagram?

a
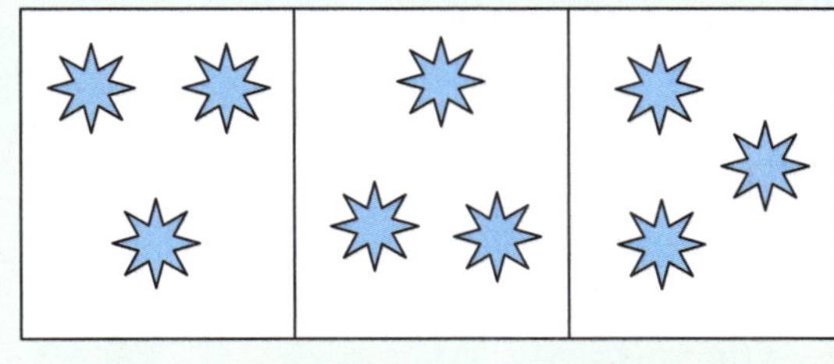

[ ] ÷ [ ] = [ ]

b
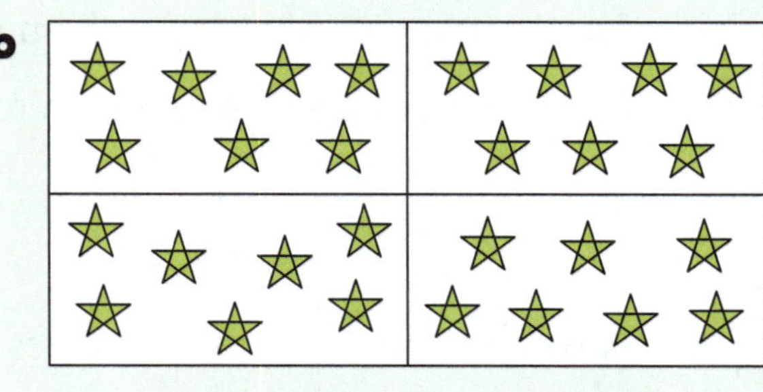

[ ] ÷ [ ] = [ ]

2 Draw a diagram to show the division and find the answer.

a $20 \div 4 =$ _______

b $18 \div 3 =$ _______

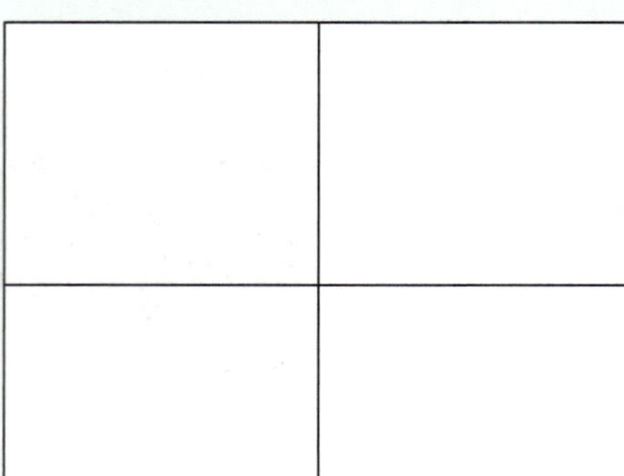

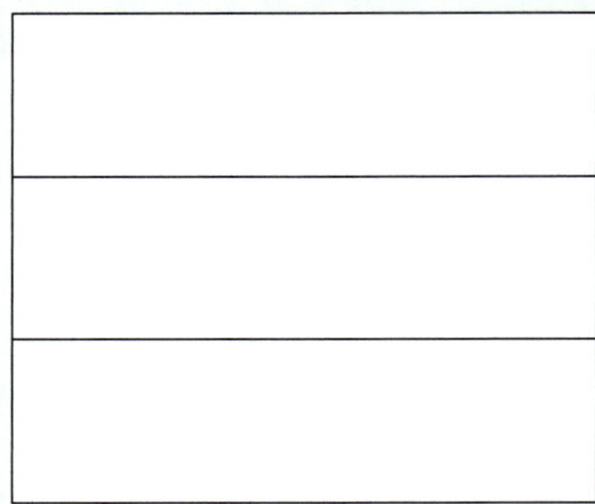

3 Use each array to write a multiplication and a division:

a 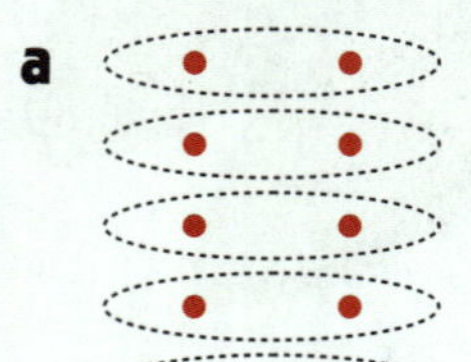

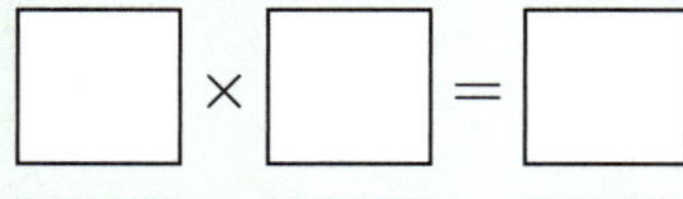

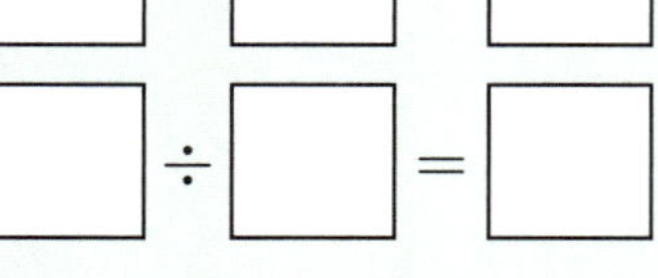

b 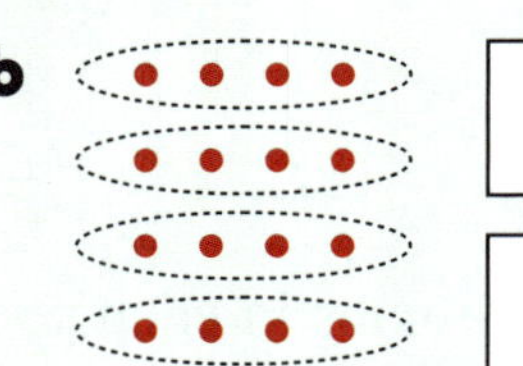

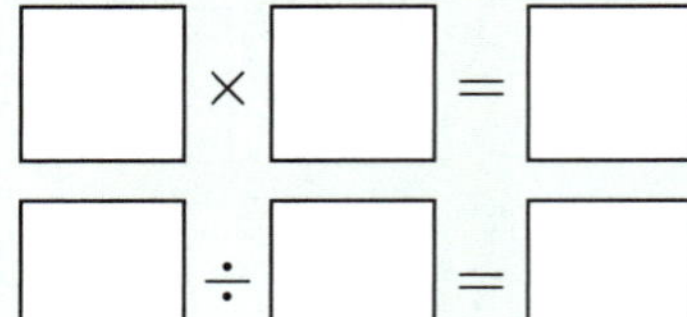

4 Solve each division:

a $21 \div 3 =$ _______

b $45 \div 9 =$ _______

c $80 \div 10 =$ _______

d $7 \div 7 =$ _______

e $36 \div 6 =$ _______

f $64 \div 8 =$ _______

5 Complete these related multiplications and divisions.

a
$4 \times 8 =$ _______

$8 \times 4 =$ _______

_______ $\div 4 = 8$

_______ $\div 8 = 4$

b
$6 \times 9 =$ _______

$9 \times 6 =$ _______

_______ $\div 6 =$ _______

_______ $\div 9 =$ _______

6 Lidia has 42 oranges. She shares them equally between 7 people.

How many oranges does each person receive?

$$\boxed{} \div \boxed{} = \boxed{}$$

Each person receives $\boxed{}$ oranges.

CHAPTER 12: FRACTIONS

Josh cut his pizza into 2 equal pieces.

He has *divided* his pizza into 2.

Each piece is called **one half** of the pizza.

The whole pizza has been divided into 2 halves.

Exercise 1

Shade one half of each shape:

a 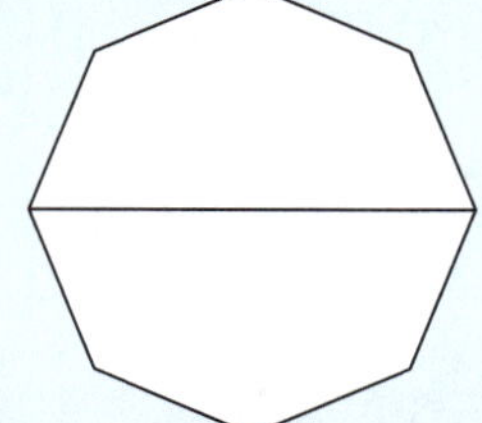   b 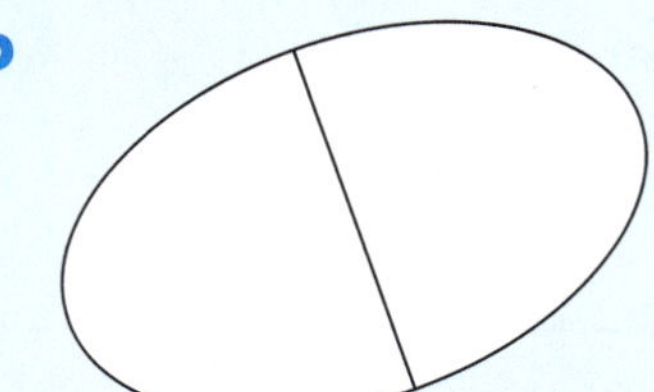   c 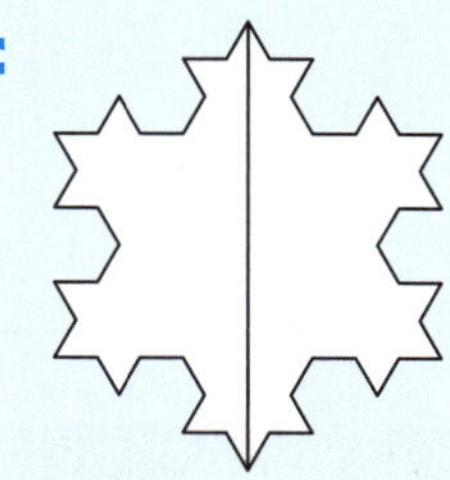

Exercise 2

✓ each shape that *is* showing one half.   ✗ each shape that *is not* showing one half.

a 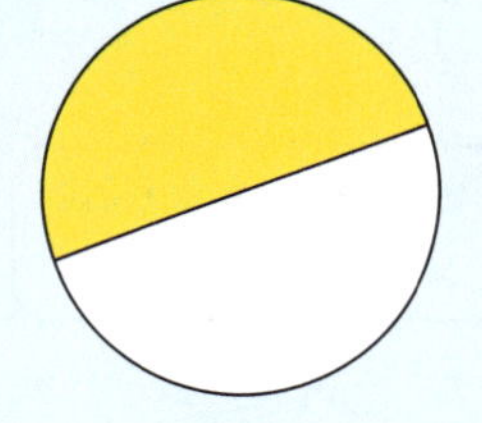   b 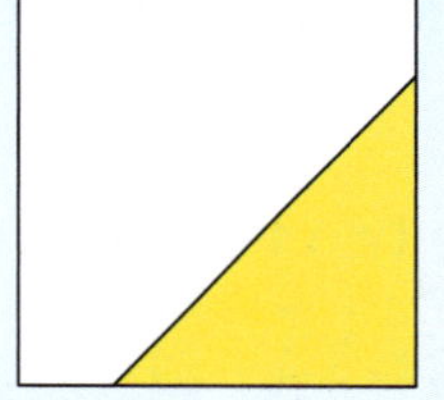   c 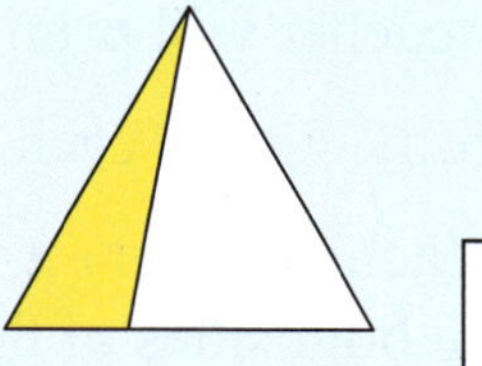

d 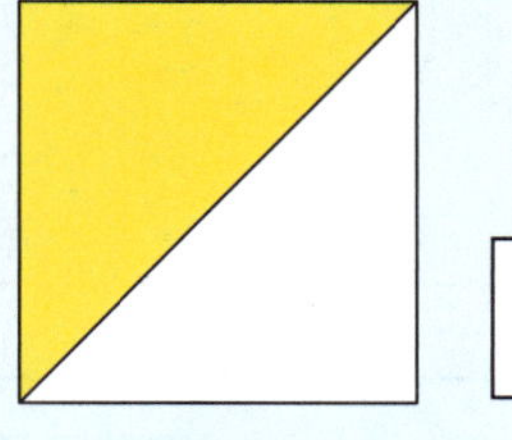   e 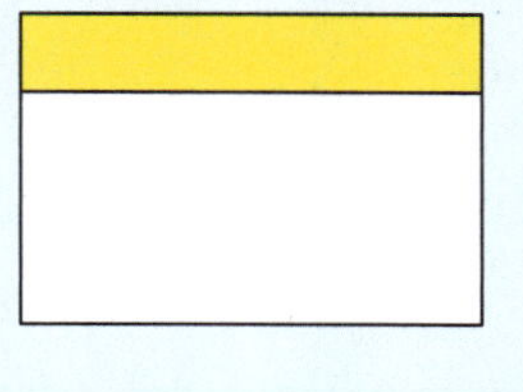   f 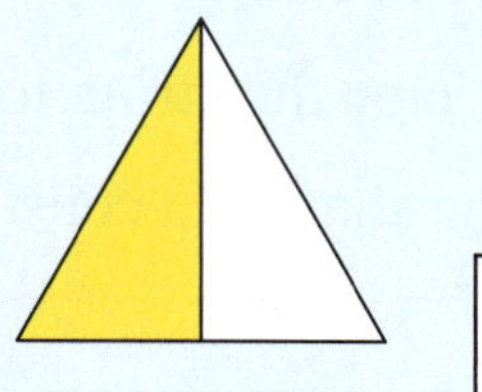

g    h 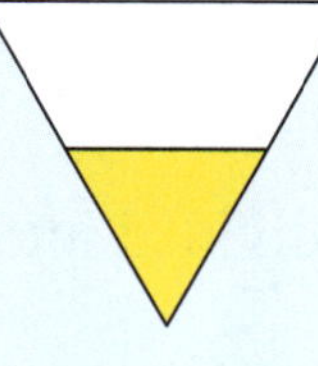   i 

Exercise 3

Shade one half of each shape:

a

b
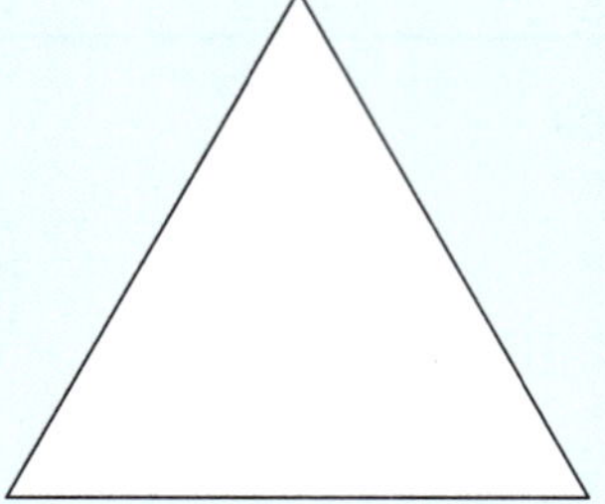

c
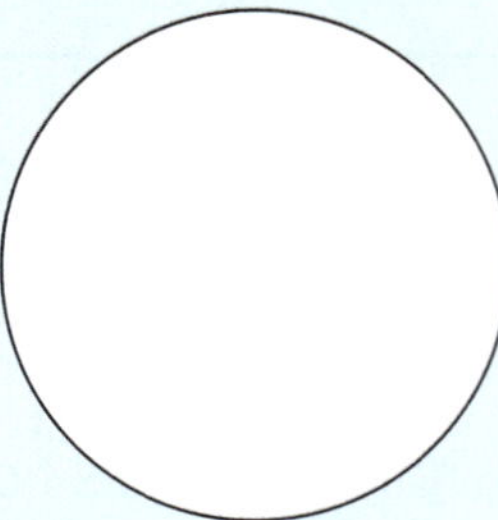

d

e
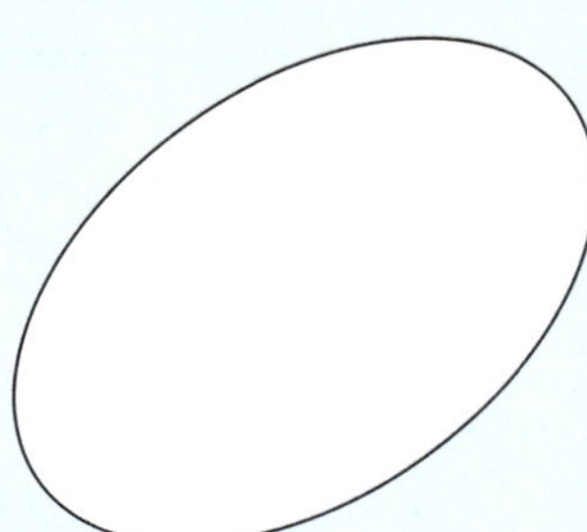

f

Activity   Showing halves

You will need:

printed sheets, scissors, coloured pencils, hole punch, string, tape

What to do:

1   Your teacher will print out a sheet of shapes to use.

2   Carefully cut out each shape.

3   Fold each shape so it is divided into two halves. Colour one half on both sides of the paper.

4   Punch a hole into one corner of each shape and thread through a piece of string.

5   Tie or tape the string to the shape.

6   Use the shapes to construct a class display or individual mobile.

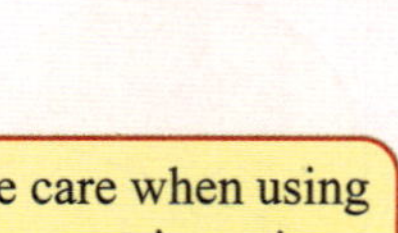

Instead of writing "one half" in words, we can write the **fraction** $\frac{1}{2}$.

numerator ⟶ 1 ⟵ We are looking at 1 part.

bar ⟶

2 ⟵ The whole is divided into 2 equal parts.

denominator

Exercise 4

Practise your spelling.

half   halves   fraction

_______________   _______________   _______________

numerator   denominator

_______________   _______________

To find **one half** of a number, we divide it into **two equal groups**.

$\frac{1}{2}$ of 16 is 8.

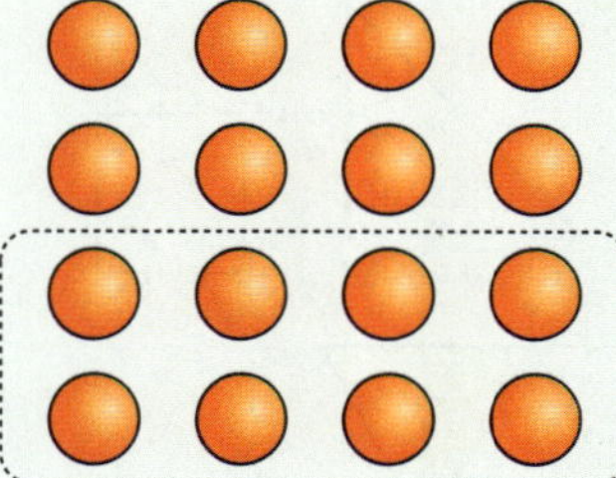

Exercise 5

Circle $\frac{1}{2}$ of the objects, and complete the statements.

a

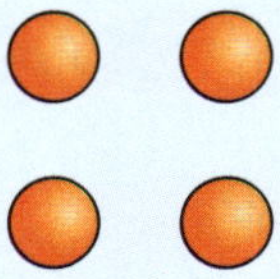

$\frac{1}{2}$ of 4 is ☐.

b

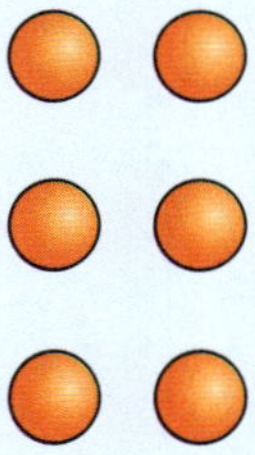

$\frac{1}{2}$ of 6 is ☐.

c

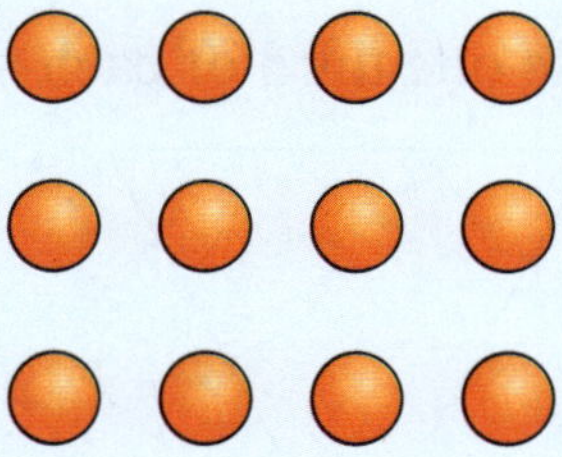

$\frac{1}{2}$ of 12 is ☐.

d

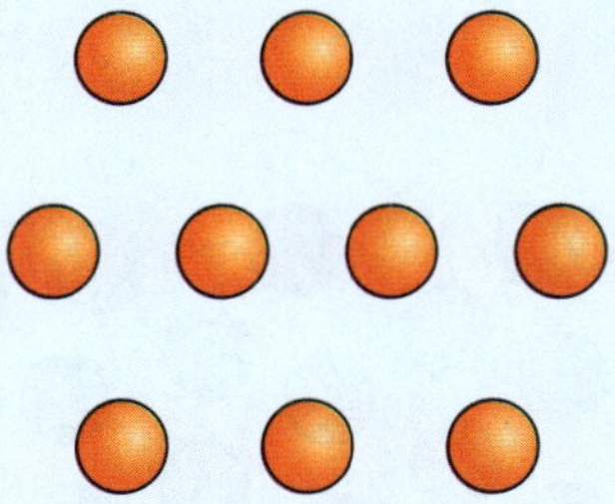

$\frac{1}{2}$ of ☐ is ☐.

e

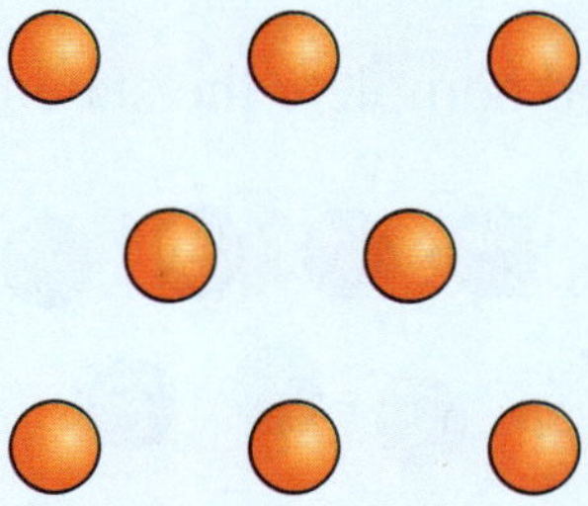

$\frac{1}{2}$ of ☐ is ☐.

f

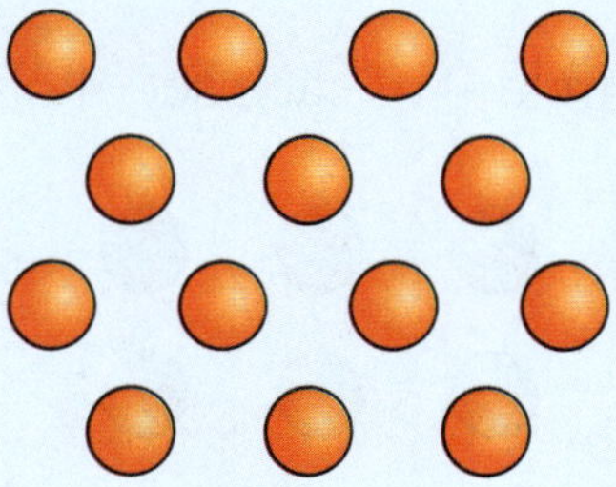

$\frac{1}{2}$ of ☐ is ☐.

Jasmine cut her pie into 3 equal pieces.

She has *divided* her pie into 3.

Each piece is called **one third** of the pie.

We can write "one third" as the fraction $\frac{1}{3}$.

1 ← We are looking at 1 part.

3 ← The whole is divided into 3 equal parts.

Exercise 6

✓ each shape that *is* showing one third.   ✗ each shape that *is not* showing one third.

a 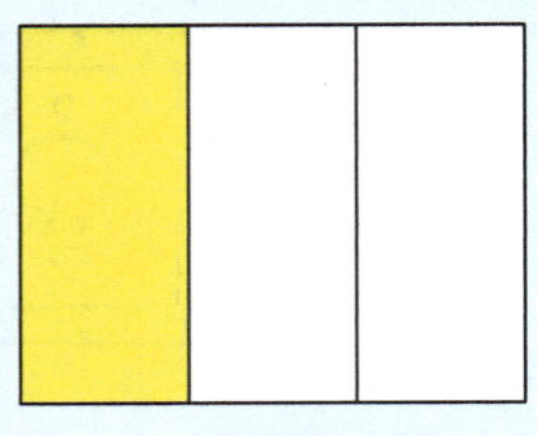

b 

c 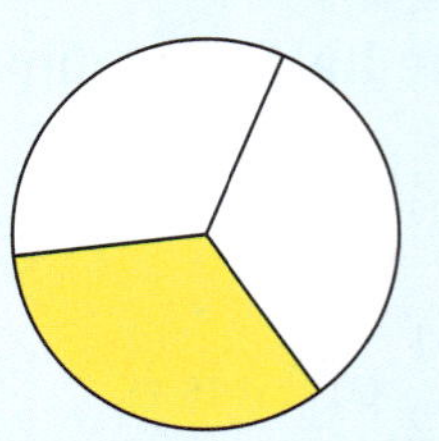

d 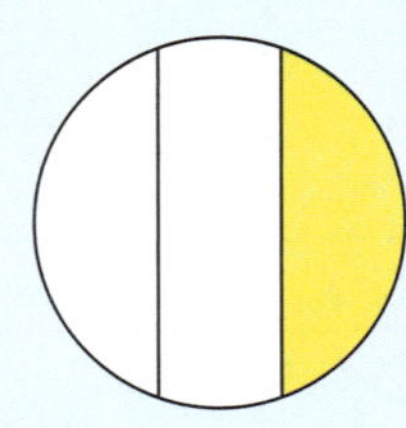

e 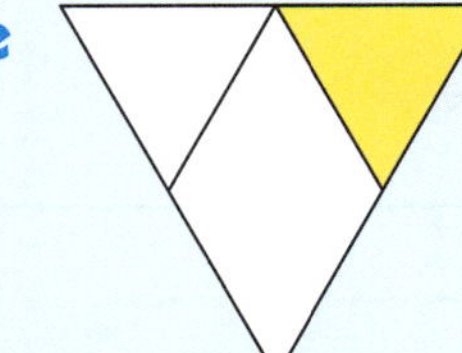 

f 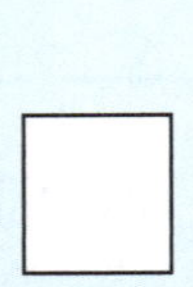

Exercise 7

Shade one third of each shape:

a 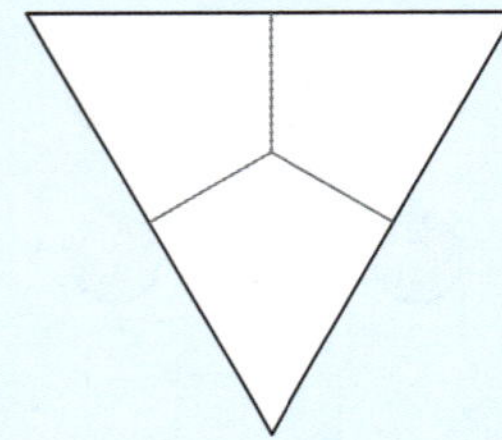

b 

c 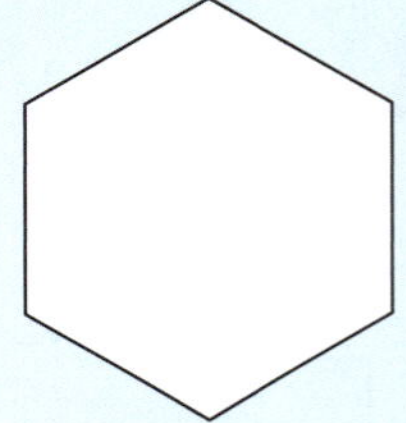

Exercise 8

Circle $\frac{1}{3}$ of each set of objects, and complete the statements:

a 

$\frac{1}{3}$ of 6 is ☐ .

b 

$\frac{1}{3}$ of ☐ is ☐ .

c 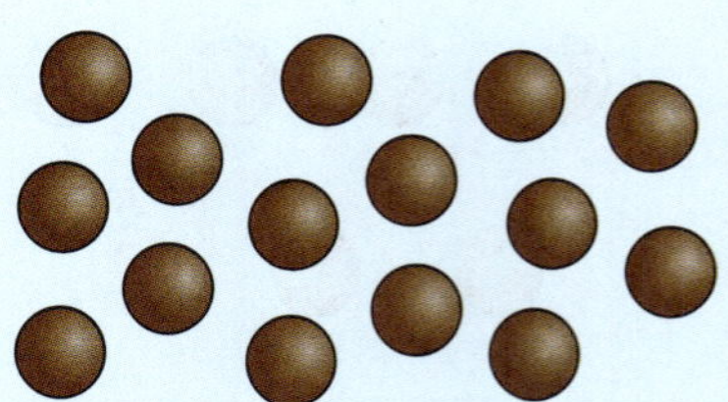

$\frac{1}{3}$ of ☐ is ☐ .

Elise cut a lamington into 4 equal pieces.

Each piece is called **one quarter** of the lamington.

We can write "one quarter" as the fraction $\frac{1}{4}$.

One whole is divided into 4 equal pieces.

Exercise 9

Shade one quarter of each shape:

a
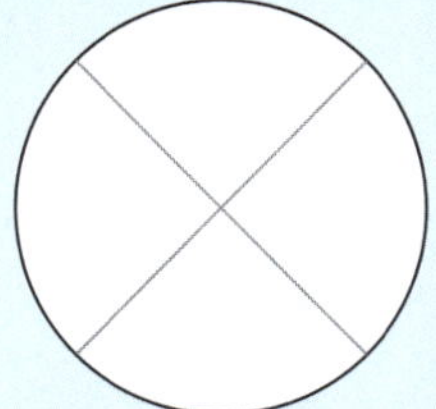

b

c
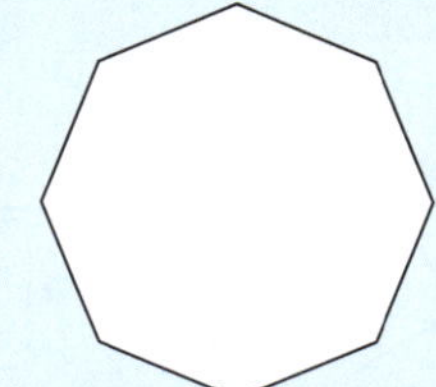

Exercise 10

Circle $\frac{1}{4}$ of each set of objects, and complete the statements:

a

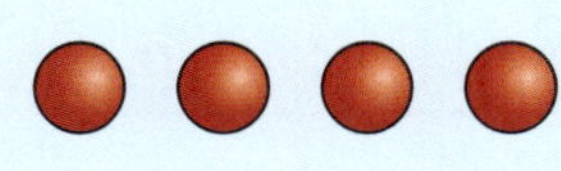

$\frac{1}{4}$ of 4 is ☐ .

b

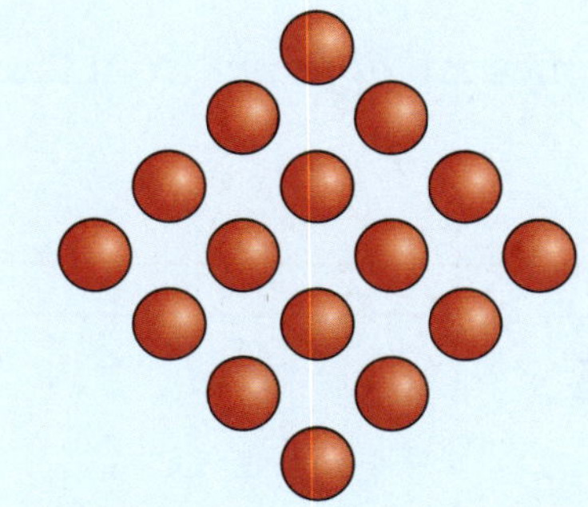

$\frac{1}{4}$ of ☐ is ☐ .

c

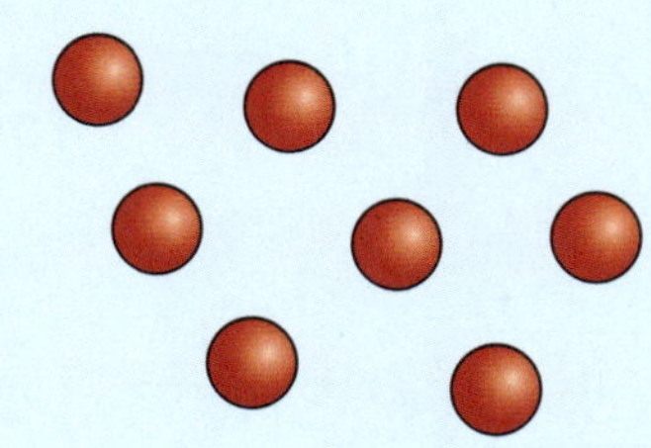

$\frac{1}{4}$ of ☐ is ☐ .

d

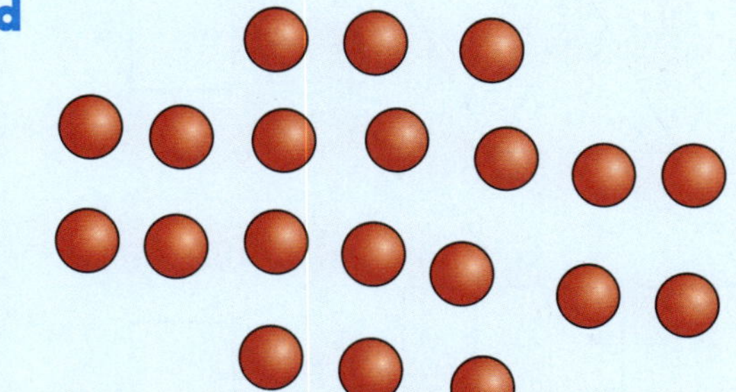

$\frac{1}{4}$ of ☐ is ☐ .

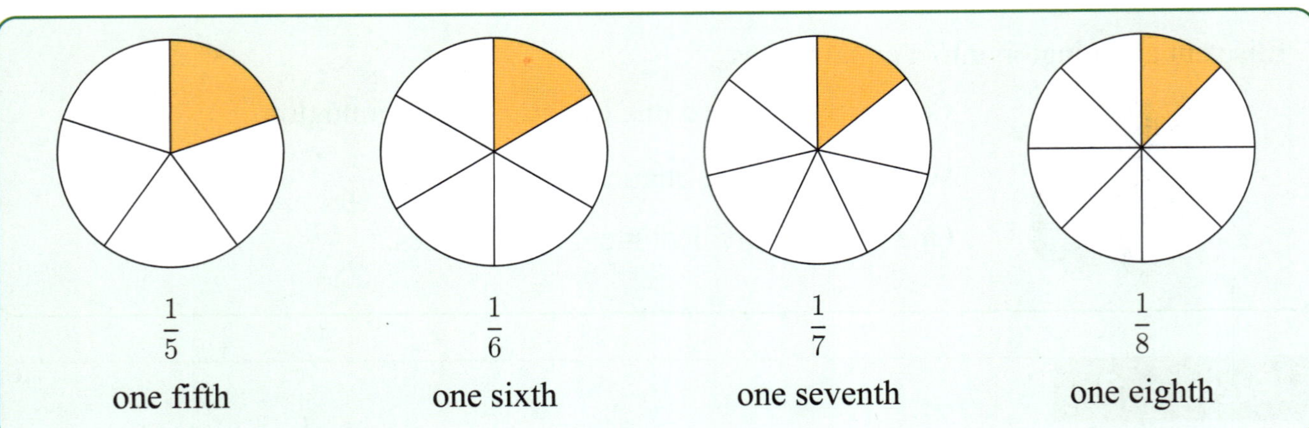

$$\frac{1}{5}$$
one fifth

$$\frac{1}{6}$$
one sixth

$$\frac{1}{7}$$
one seventh

$$\frac{1}{8}$$
one eighth

Exercise 11

$$\dfrac{1}{\boxed{}}$$ ← The whole is divided into _______ equal parts.

one _________________

Discussion

What set of numbers are the fraction names similar to?   _______________________________

What fraction names are *different* from these numbers?

_______________________________   _______________________________

Exercise 12

Name the fraction.

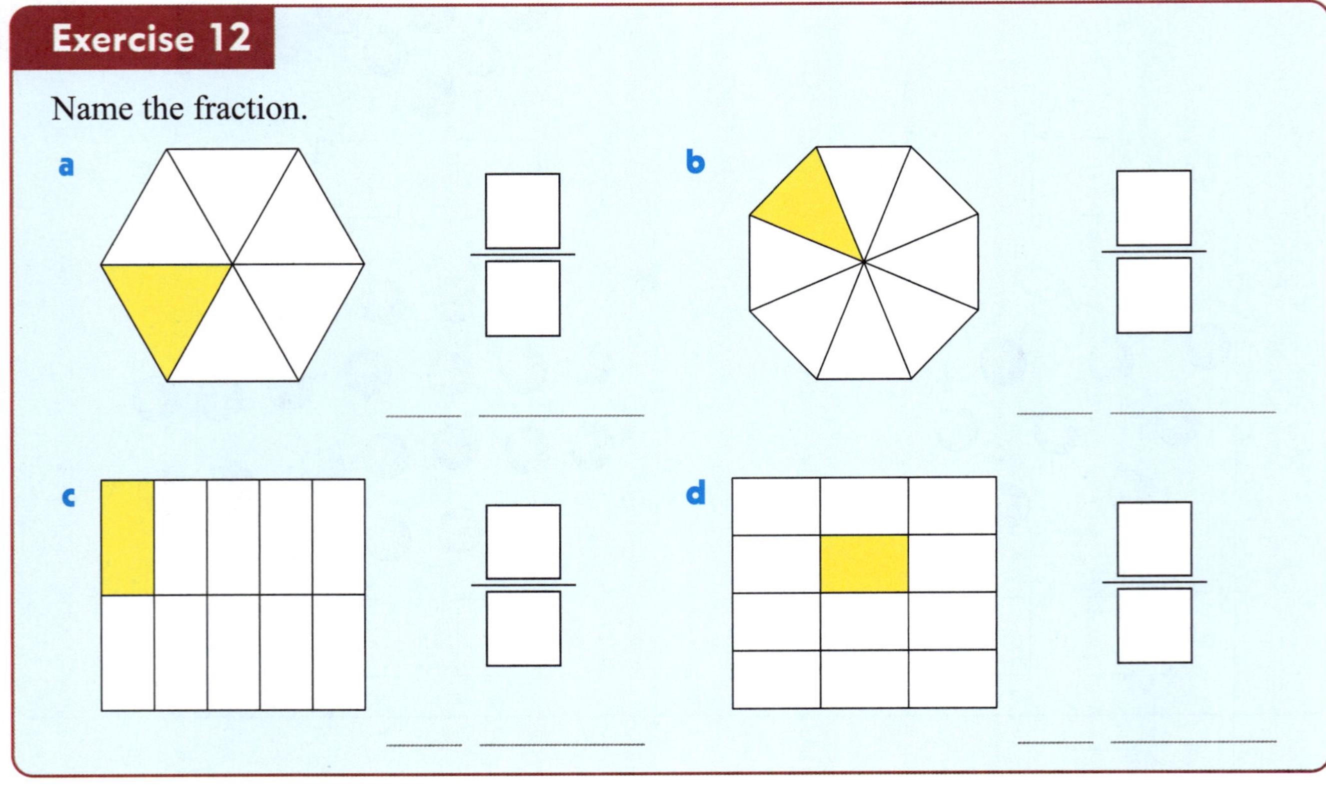

a

b

c

d

Exercise 13

a Shade $\frac{1}{2}$.

b Shade $\frac{1}{3}$.

c Shade $\frac{1}{4}$.

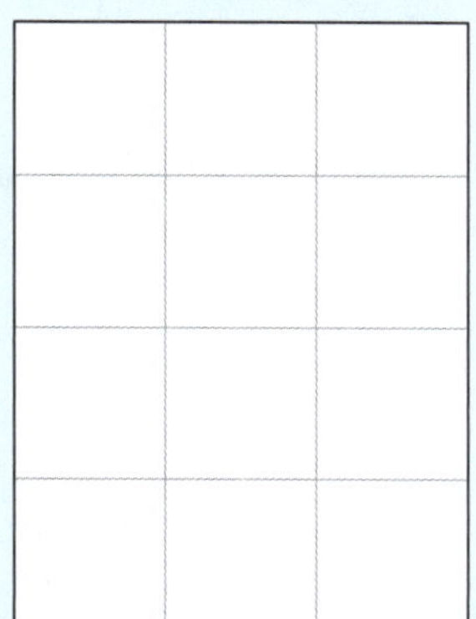

Exercise 14

a Sally has 32 beads to use to make 4 bracelets. She puts the same number on each.

Sally puts $\dfrac{\boxed{}}{\boxed{}}$ of the beads on each bracelet.

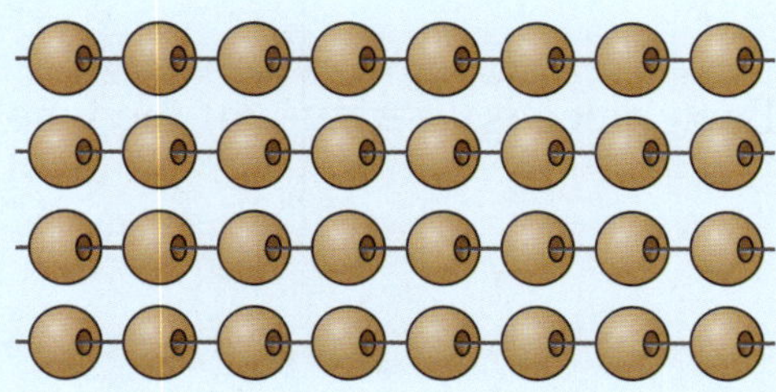

$\dfrac{\boxed{}}{\boxed{}}$ of 32 is $\boxed{}$.

She puts $\boxed{}$ beads on each bracelet.

b Benny has 27 superhero figurines. He shares them equally between 3 friends so they can save Earth from alien invaders.

Each person receives $\dfrac{\boxed{}}{\boxed{}}$ of the figurines.

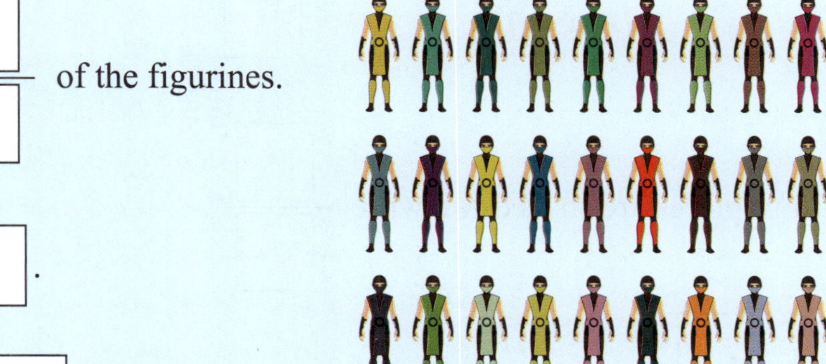

$\dfrac{\boxed{}}{\boxed{}}$ of 27 is $\boxed{}$.

So, each person receives $\boxed{}$ figurines.

This whole is divided into 3 equal pieces.

Each piece is $\frac{1}{3}$ of the whole.

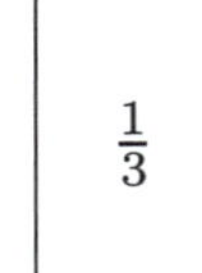

If two pieces are shaded, we write $\frac{2}{3}$.

If three pieces are shaded, we write $\frac{3}{3}$.

$$\frac{3}{3} = 1$$

Exercise 15

This whole is divided into ______ equal pieces.

Each piece is $\dfrac{\square}{\square}$ of the whole.

If two pieces are shaded, we write $\dfrac{\square}{\square}$.

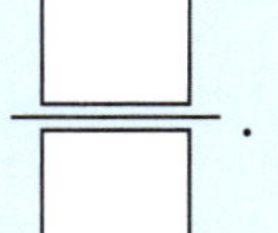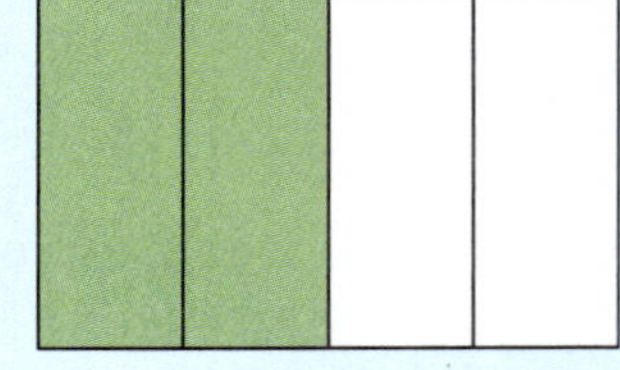

If three pieces are shaded, we write $\dfrac{\square}{\square}$.

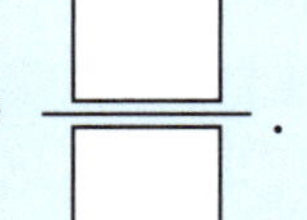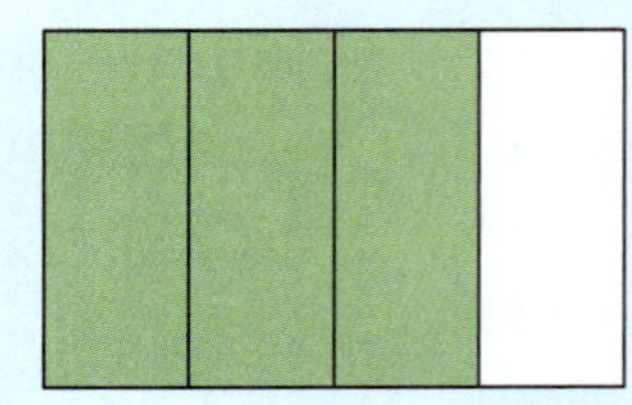

If four pieces are shaded, we write $\dfrac{\square}{\square}$.

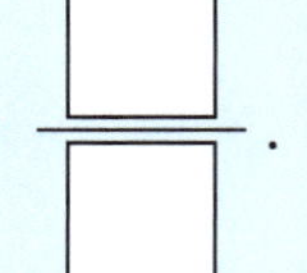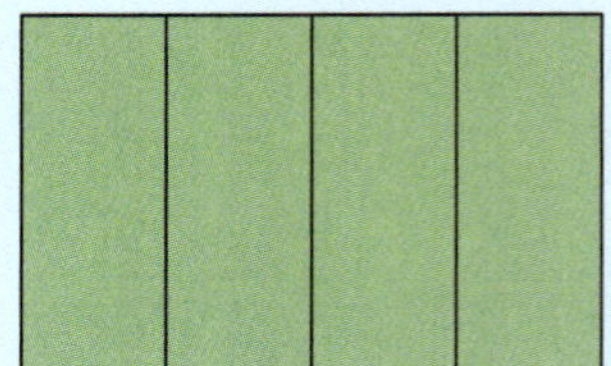

$$\dfrac{\square}{\square} = 1$$

Exercise 16

What fraction is shaded?

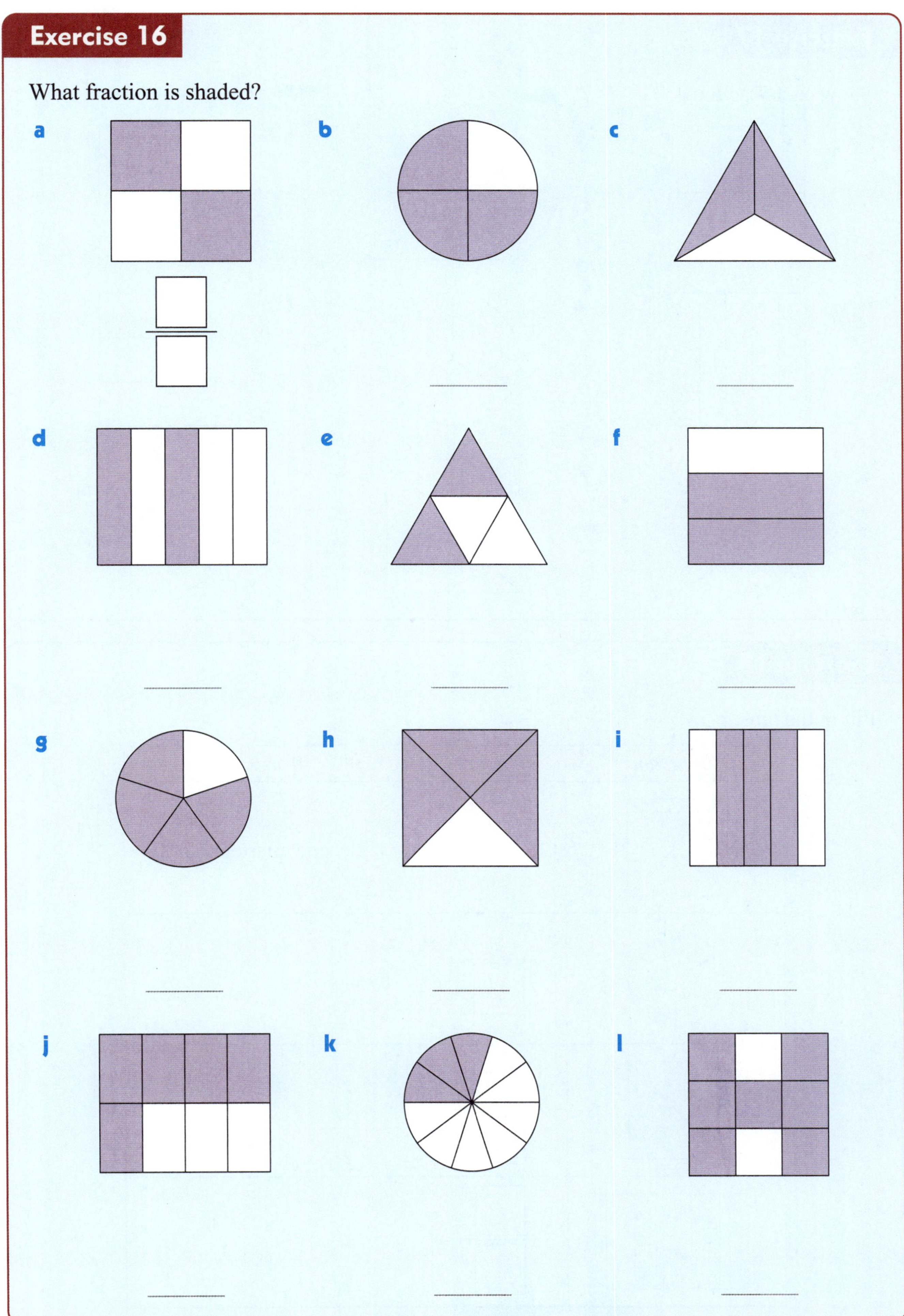

Exercise 17

Show each fraction:

a $\dfrac{3}{4}$ 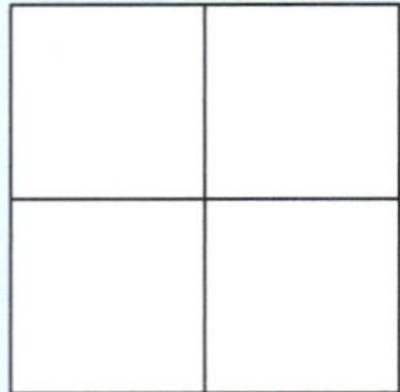

b $\dfrac{2}{3}$ 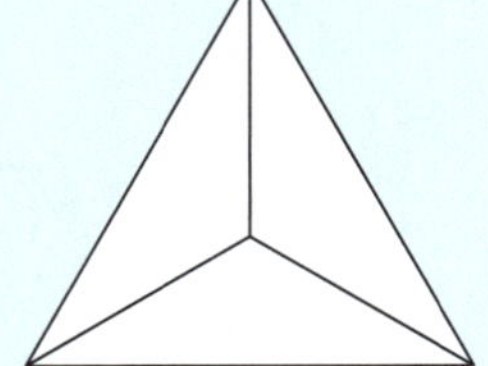

c $\dfrac{1}{4}$ 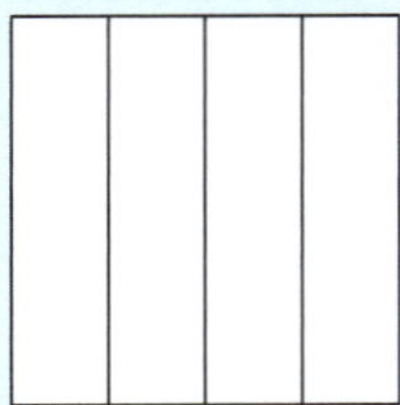

d $\dfrac{2}{5}$ 

e $\dfrac{2}{3}$ 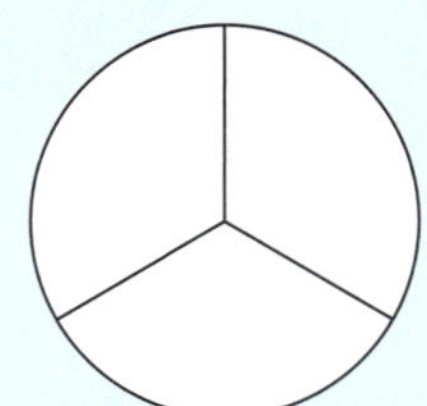

f $\dfrac{5}{6}$ 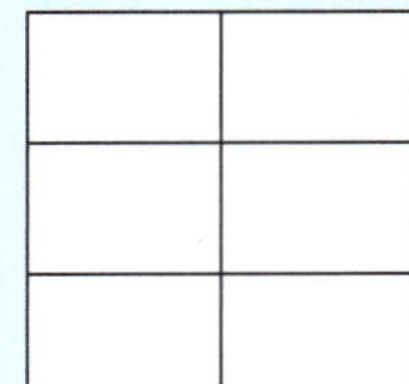

g $\dfrac{4}{5}$ 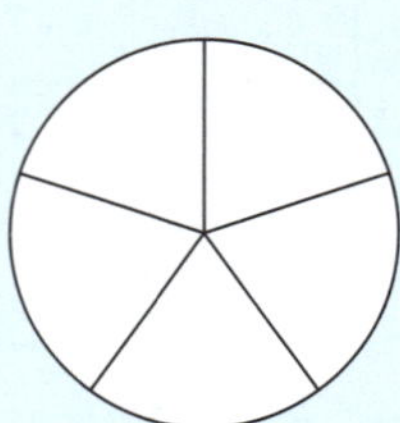

h $\dfrac{2}{4}$ 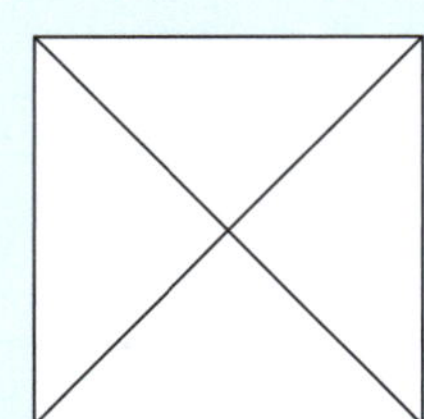

i $\dfrac{3}{8}$ 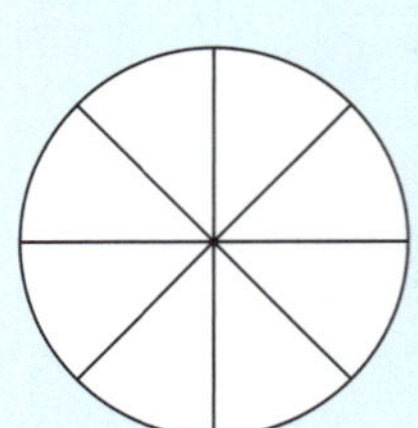

Exercise 18

Fill in the table.

Diagram	Fraction	Words
	$\dfrac{1}{4}$	one quarter
	$\dfrac{2}{\Box}$	two _________
	$\dfrac{\Box}{\Box}$	_________
	$\dfrac{\Box}{\Box}$	_________

Exercise 19

Shade each fraction:

a $\dfrac{2}{4}$ 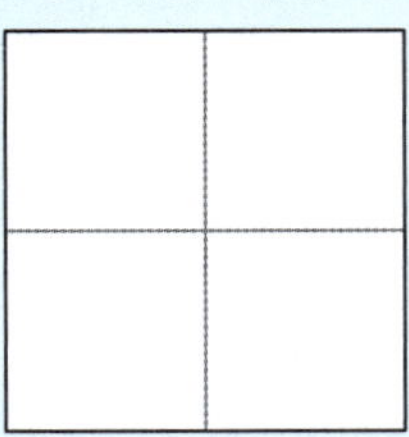

b $\dfrac{1}{2}$ 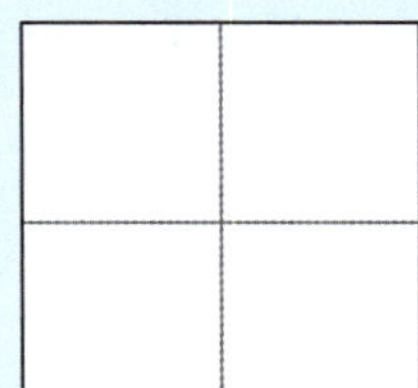

What can you say about $\dfrac{2}{4}$ and $\dfrac{1}{2}$?

Exercise 20

Show each fraction:

a $\dfrac{2}{3}$ 

b $\dfrac{4}{6}$ 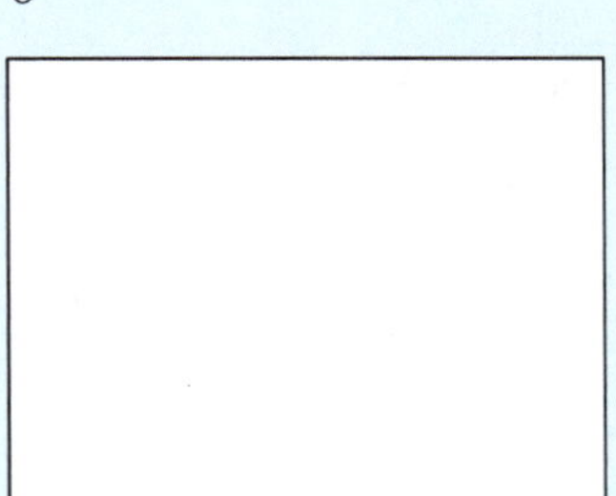

c three thirds 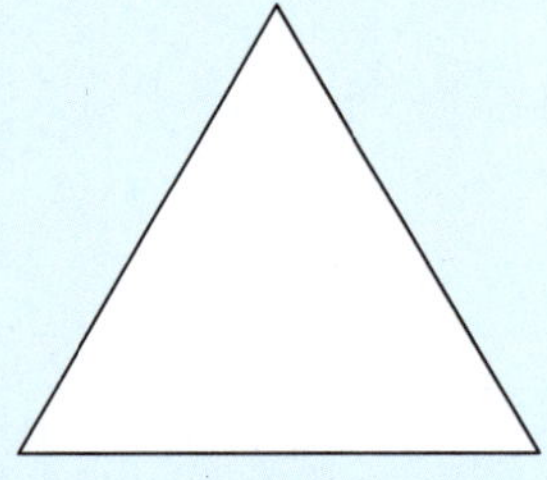

d one fifth 

e three quarters 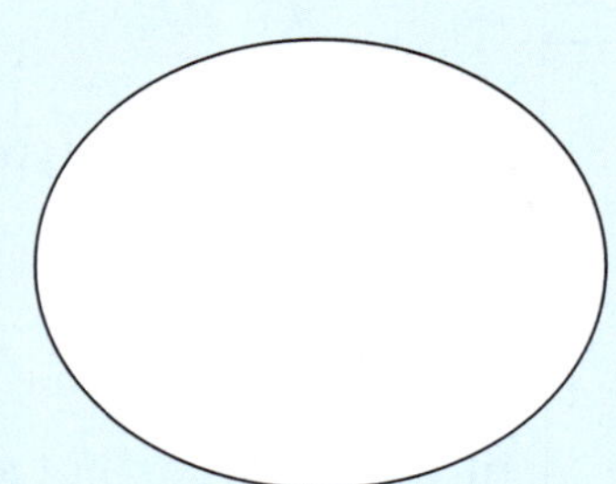

f $\dfrac{3}{5}$ 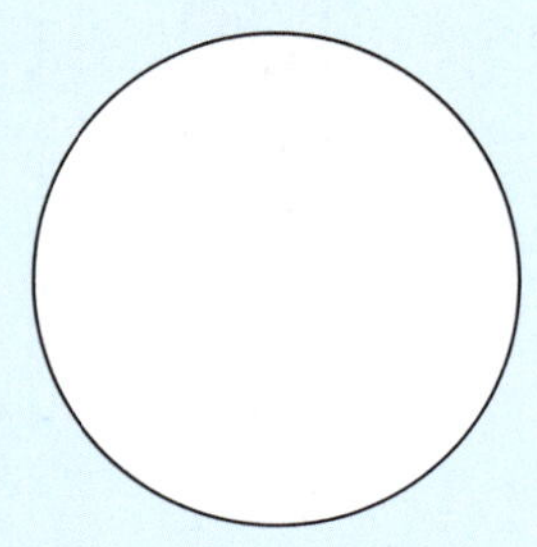

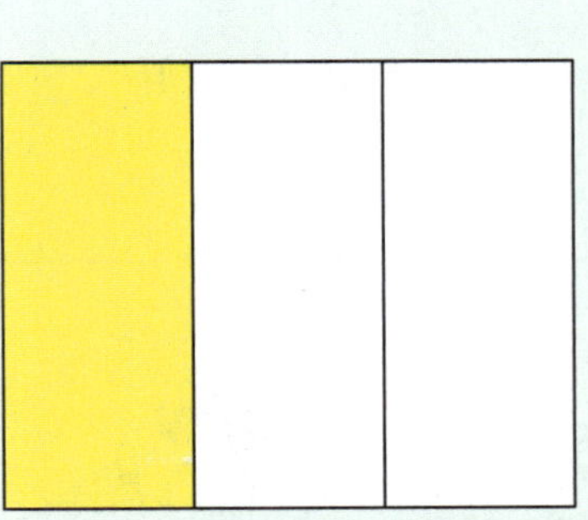

Brayden has coloured $\dfrac{1}{3}$ of this rectangle.

$\dfrac{2}{3}$ of the rectangle is not coloured.

$\dfrac{1}{3} + \dfrac{2}{3} = \dfrac{3}{3}$ which is one whole.

Exercise 21

Fill in the spaces.

a

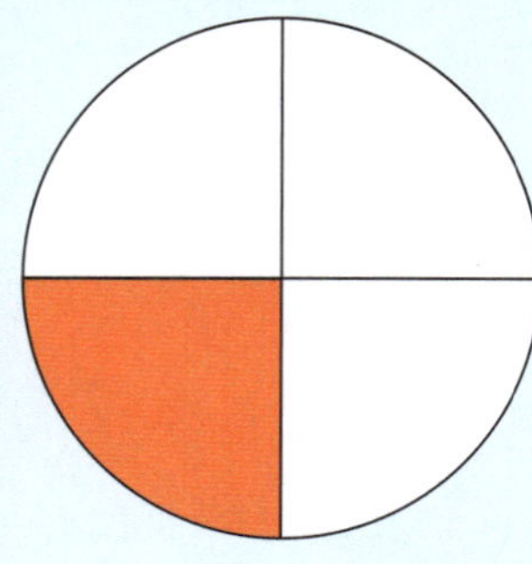

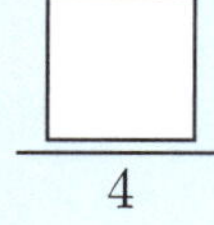 $\dfrac{}{4}$ of the circle is shaded.

$\dfrac{}{4}$ of the circle is not shaded.

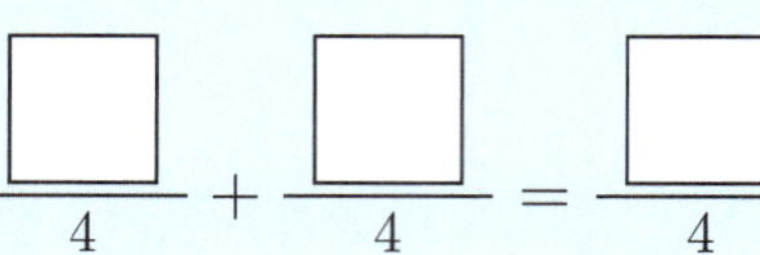 $\dfrac{}{4} + \dfrac{}{4} = \dfrac{}{4}$ which is one whole.

b

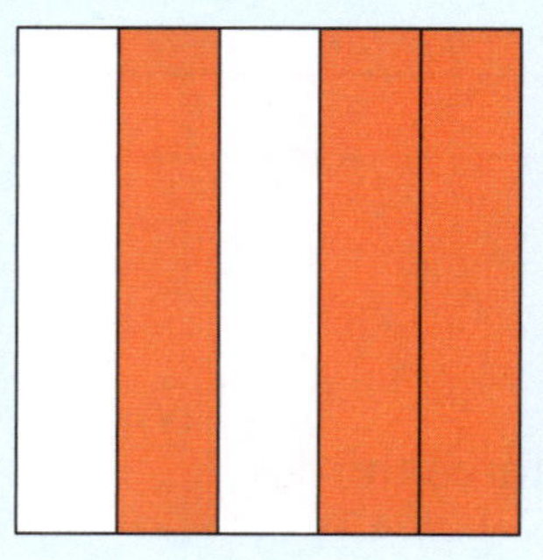

$\square$ of the square is shaded.

$\square$ of the square is not shaded.

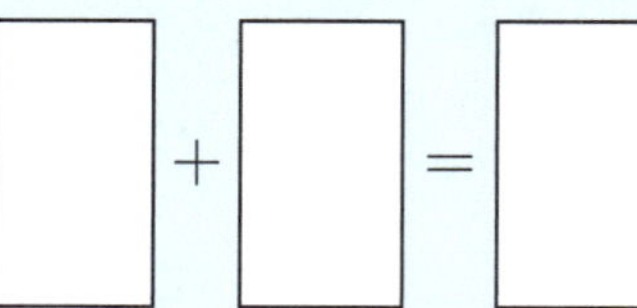 $\square + \square = \square$ which is one whole.

c

$\square$ of the square is shaded blue.

$\square$ of the square is shaded red.

$\square$ of the square is not shaded.

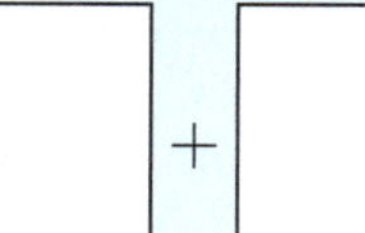 $\square + \square + \square = \square$ which is one whole.

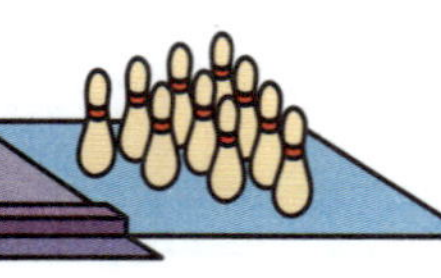

Revision

1 Draw a circle around the *denominator* of the fraction.

The denominator tells us _______________________________

____

$$\frac{2}{3}$$

2 Shade the fraction.

a one half

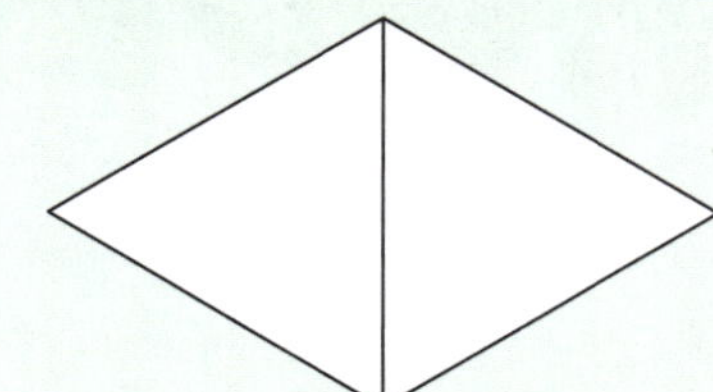

b one quarter

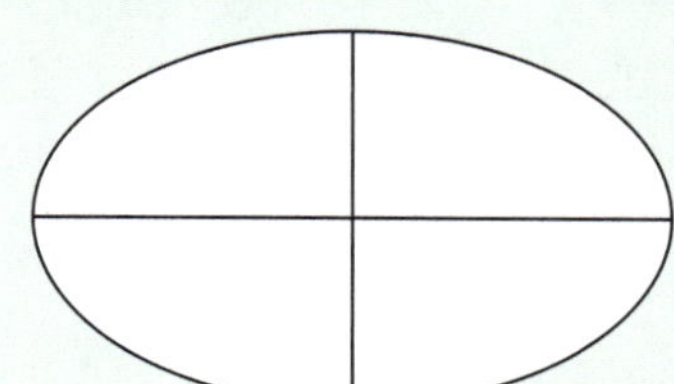

c one sixth

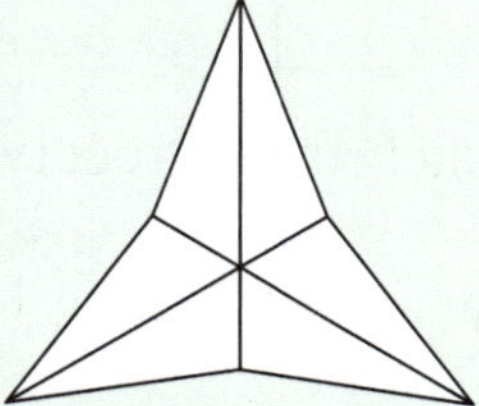

3 Circle the fraction of the objects, and complete the statements.

a

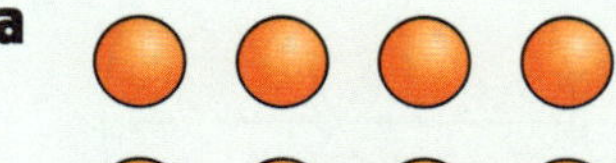
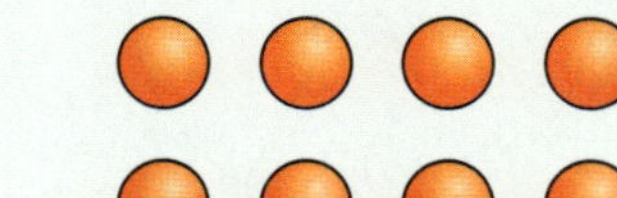

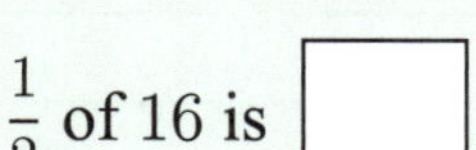

$\frac{1}{2}$ of 16 is ☐ .

b

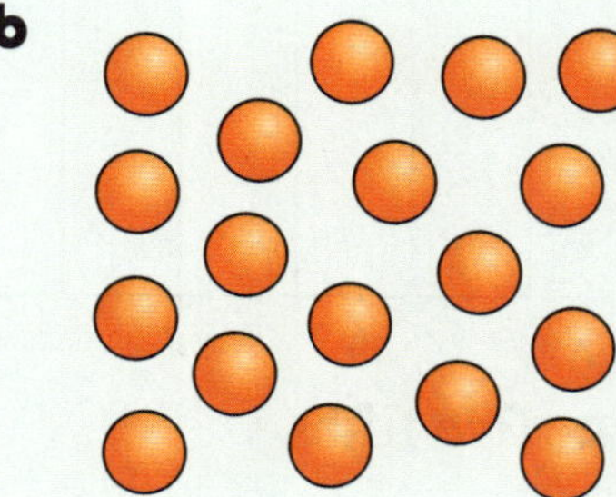

$\frac{1}{3}$ of 18 is ☐ .

c

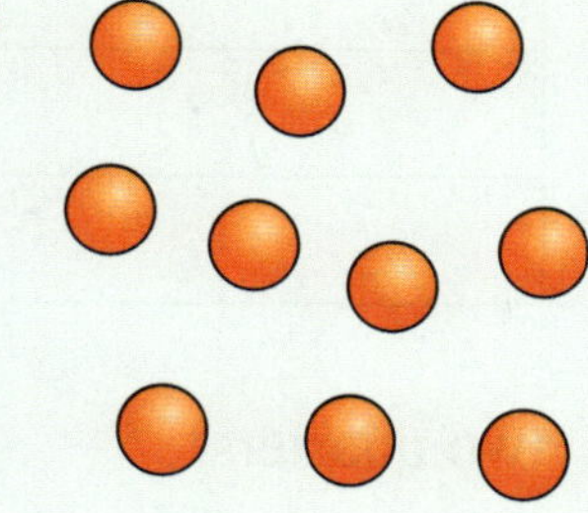

$\frac{1}{5}$ of 10 is ☐ .

4

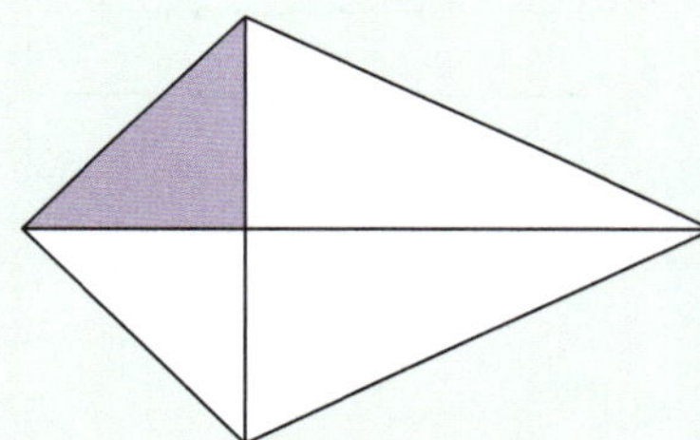

The diagram does *not* show $\frac{1}{4}$ because the 4 pieces are

not ___________________ .

5 What fraction is shaded?

a

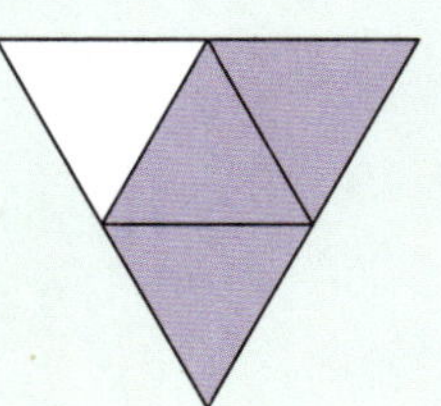

b

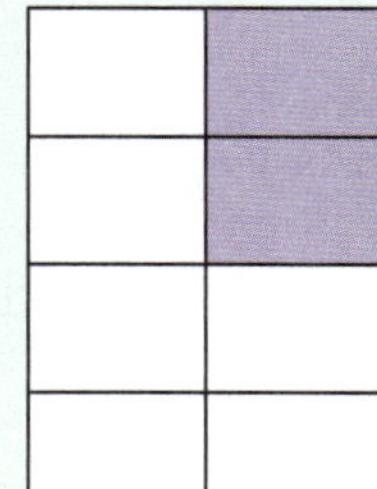

c

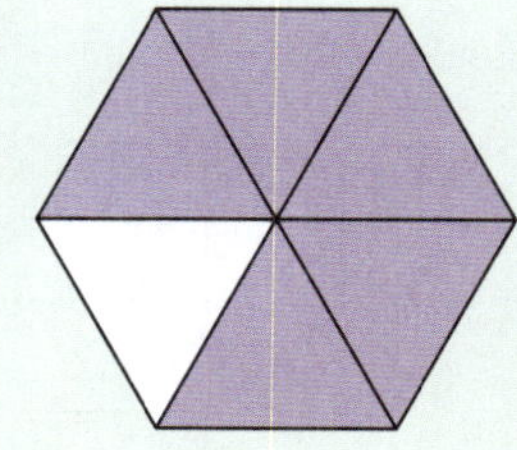

d

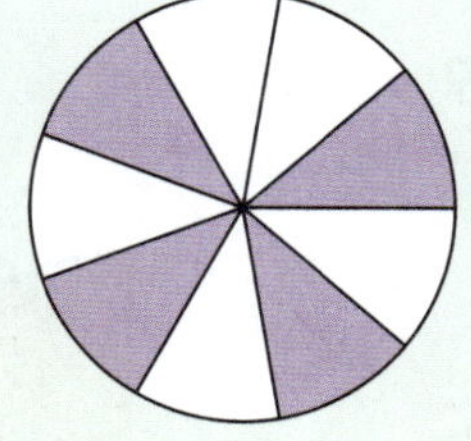

_______   _______   _______   _______

6 Di grew 25 plants in pots.

She shared them equally between 5 friends.

Each friend received $\dfrac{\boxed{}}{\boxed{}}$ of the plants.

$\dfrac{\boxed{}}{\boxed{}}$ of 25 is $\boxed{}$.

So, each friend received $\boxed{}$ plants.

7 Show each fraction:

a $\dfrac{2}{3}$

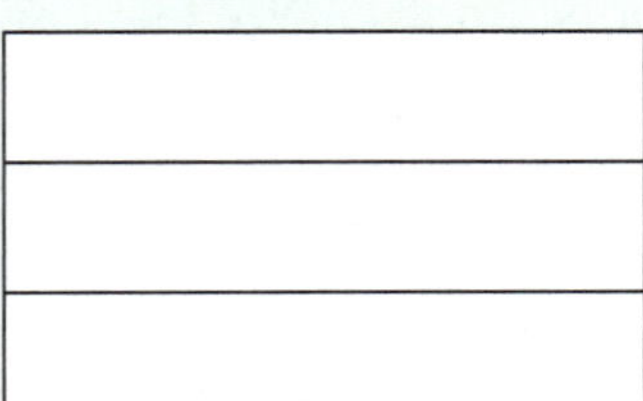

b $\dfrac{4}{5}$

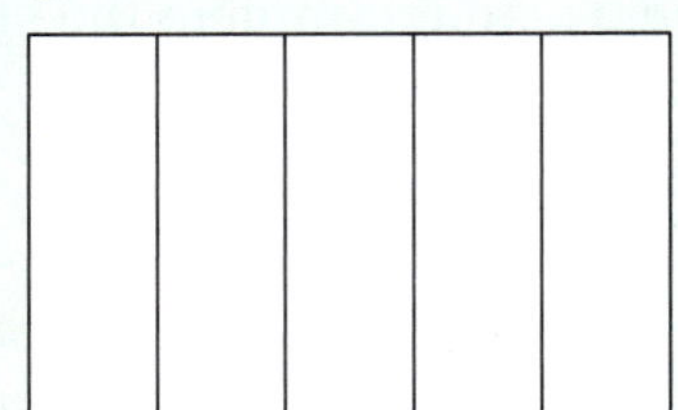

c $\dfrac{2}{6}$

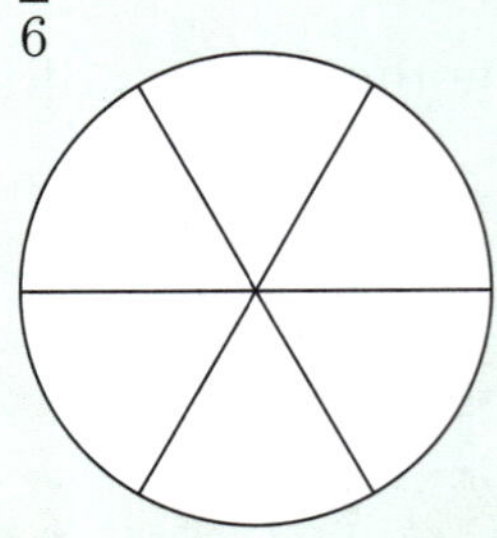

d two quarters

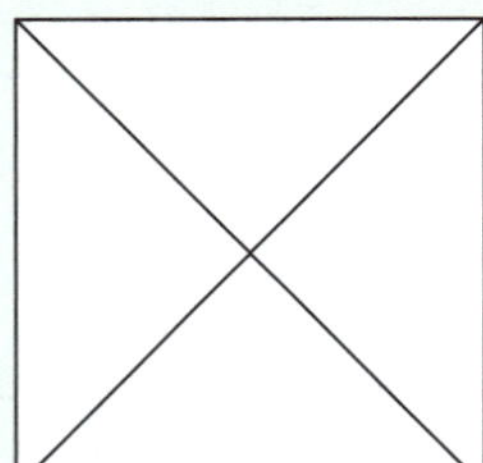

e five fifths

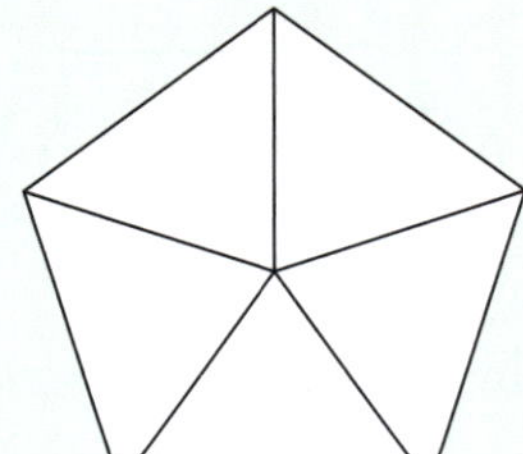

f $\dfrac{3}{5}$

8 Write the fraction in words.

a $\dfrac{2}{5}$ _______________________________

b $\dfrac{3}{3}$ _______________________________

c $\dfrac{1}{4}$ _______________________________

9 Complete:

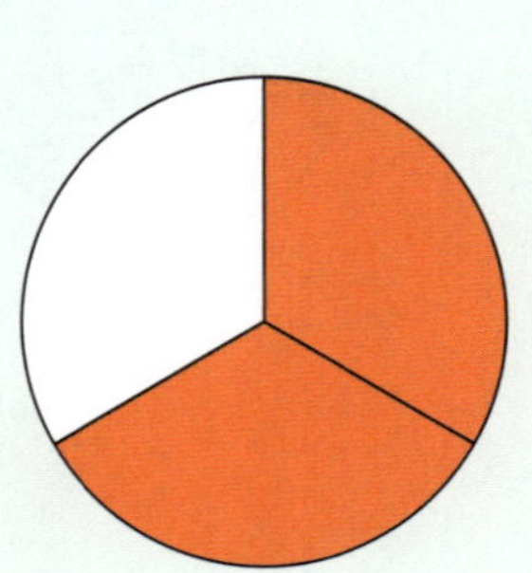

$\dfrac{\boxed{}}{3}$ of the circle is shaded.

$\dfrac{\boxed{}}{3}$ of the circle is not shaded.

$\dfrac{\boxed{}}{3} + \dfrac{\boxed{}}{3} = \dfrac{\boxed{}}{3}$ which is one whole.

These are the parts of a day:

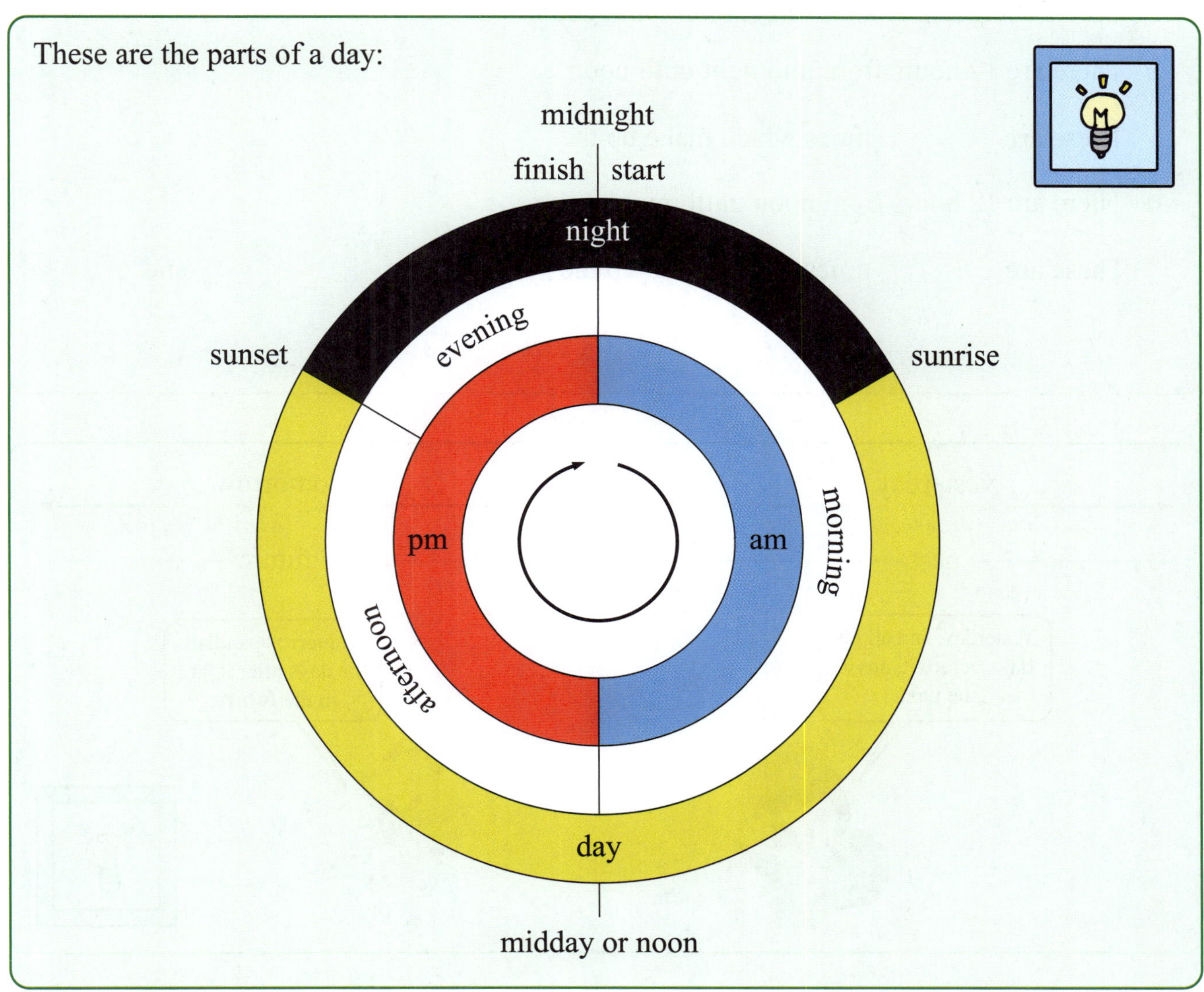

Exercise 1

Practise your writing.

midnight	__________	morning	__________
midday	__________	afternoon	__________
noon	__________	evening	__________
day	__________	sunrise	__________
night	__________	sunset	__________

Exercise 2

Each day is divided into 24 **hours**.

Complete using *am*, *pm*, *morning*, *afternoon*, and *evening*:

a There are 12 hours from midnight until noon.

These are __________ times which make up the __________________________.

b There are 12 hours from noon until midnight.

These are __________ times which make up the __________________________ and

__________________________ .

Exercise 3

Practise your writing.

yesterday __________________________   past __________________________

today __________________________   now __________________________

tomorrow __________________________   future __________________________

We use a **clock** to tell the time.

The **minute hand** is longer. It is pointing to 12.

The **hour hand** is shorter. It is pointing to 2.

The time is 2 o'clock.

The hands on a clock turn **clockwise**.

Exercise 4

Exercise 4

What time is shown on the clock?

a

__________ o'clock

b

____

c

____

The minute hand tells us where we are in the hour.

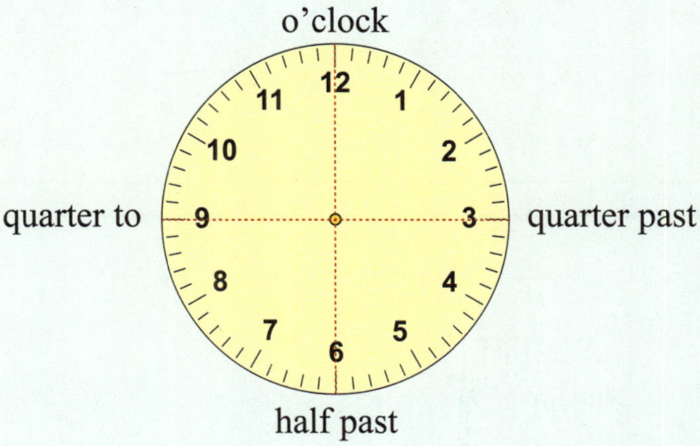

On this clock, the minute hand has turned from 12 to 3.

The minute hand is a **quarter** of the way around the clock.

The hour hand is between 5 and 6.

The time is quarter past 5.

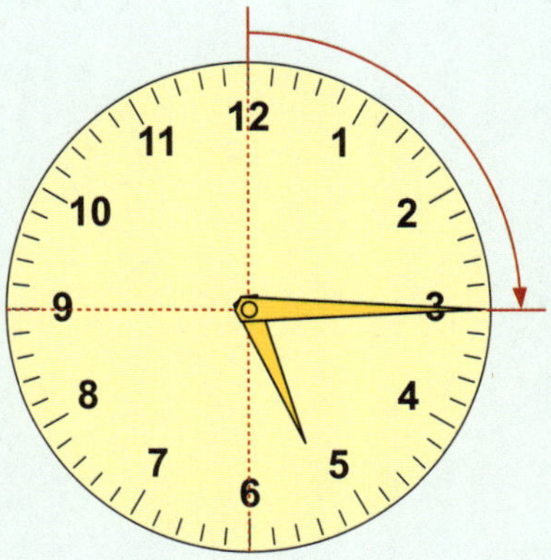

Exercise 5

What time is shown?

a

_______________ past 3

b

____

c

____

d

____

e

____

f

____

g

____

h

____

i

____

Exercise 6

Label this clock.

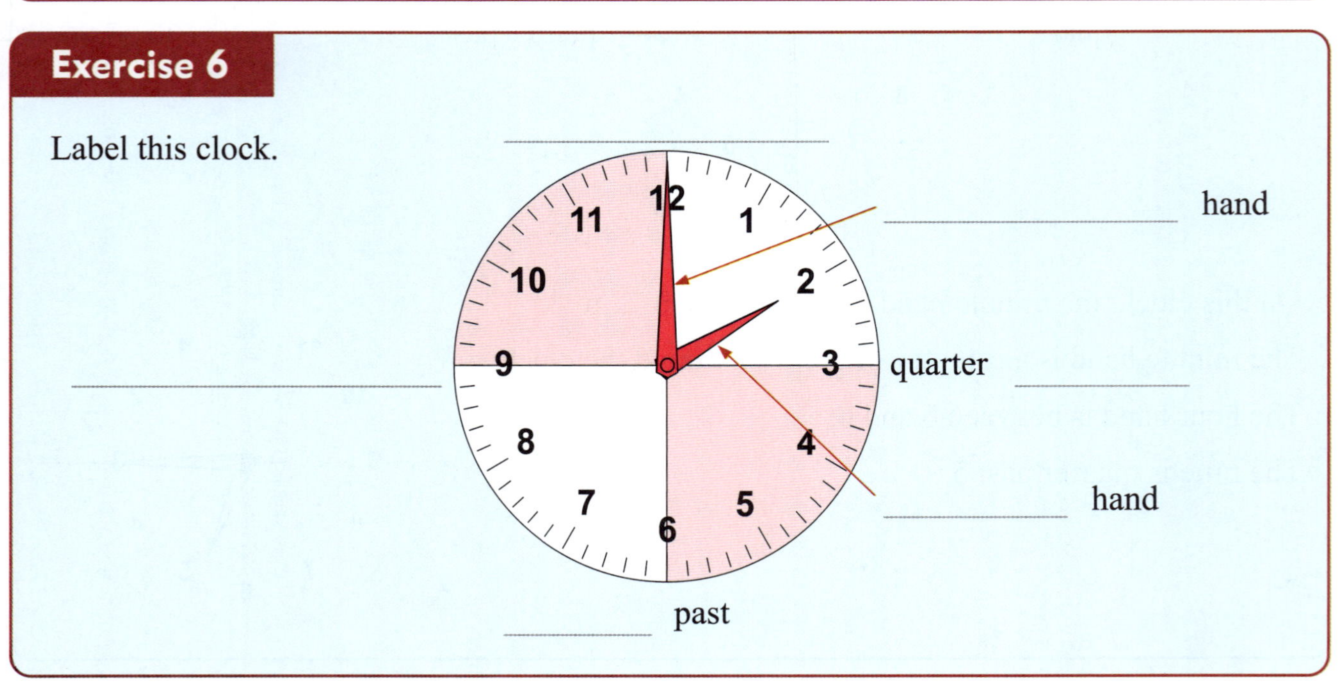

Exercise 7

Draw hands on each clock face to show the time:

a 3 o'clock

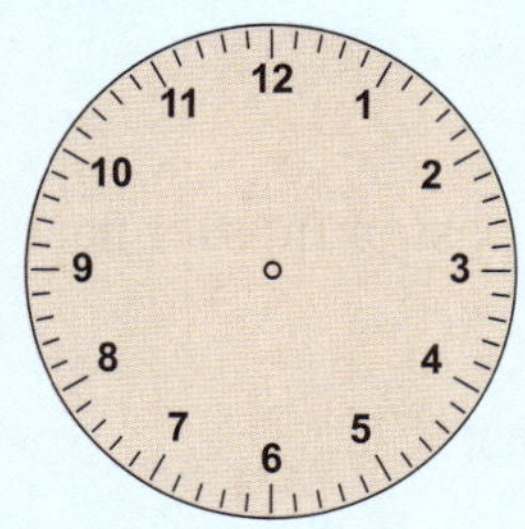

b 9 o'clock

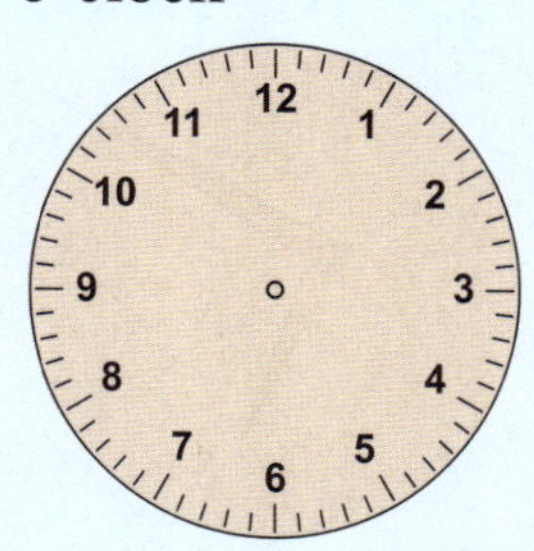

c half past 2

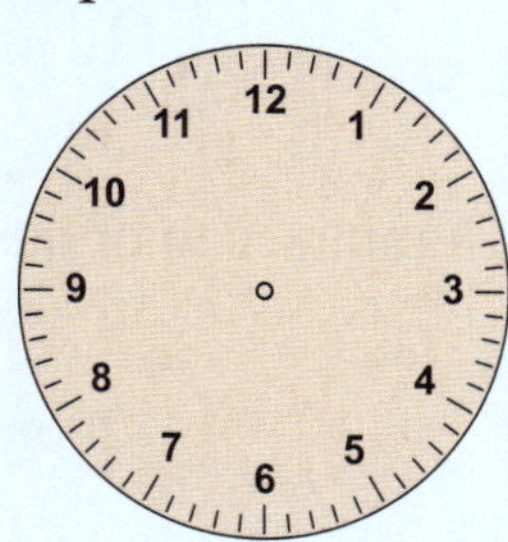

d half past 12

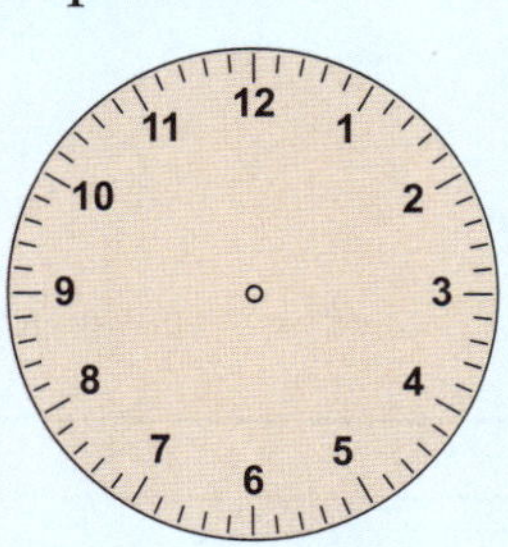

e quarter past 7

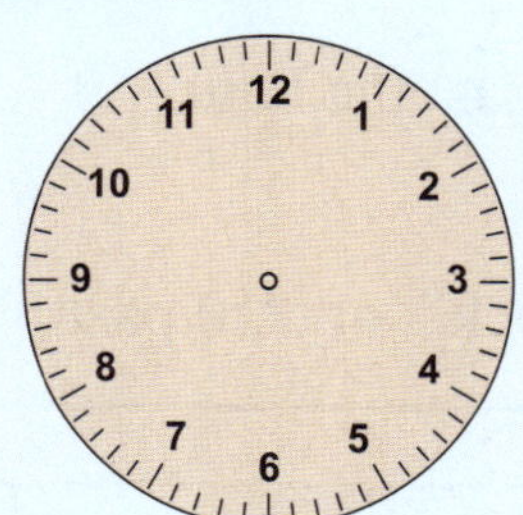

f quarter past 1

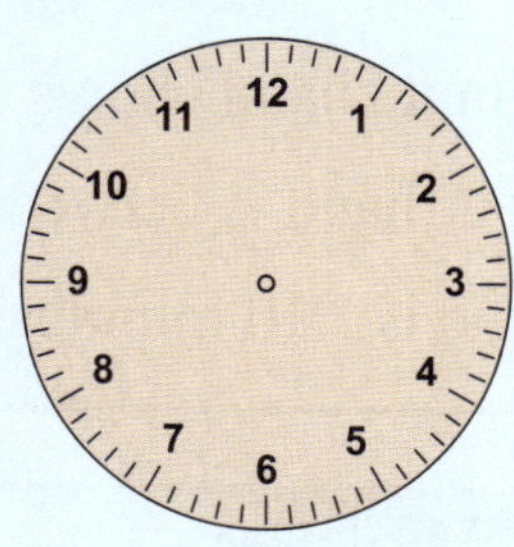

g quarter to 8

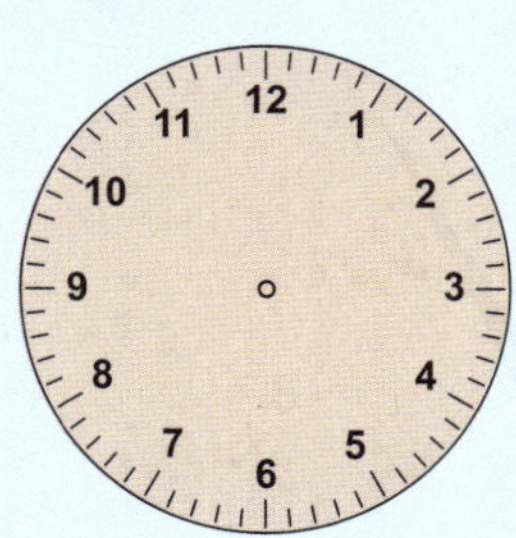

h quarter to 6

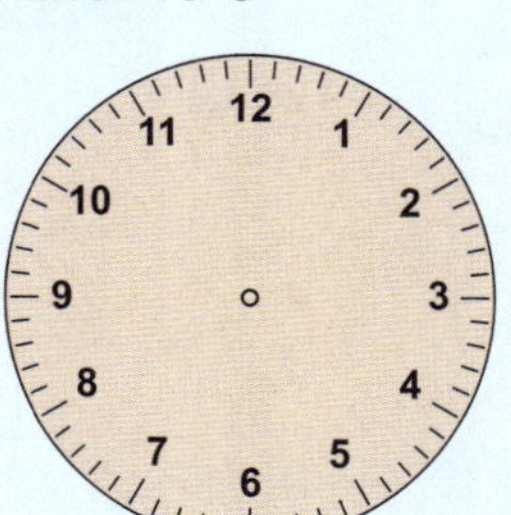

i quarter to 11

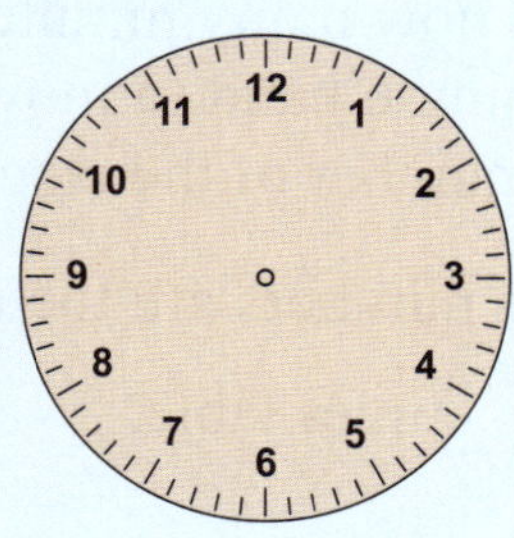

j 6 o'clock

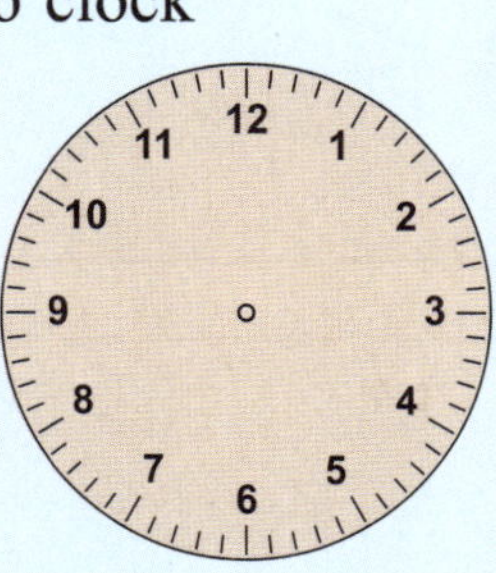

k half past 1

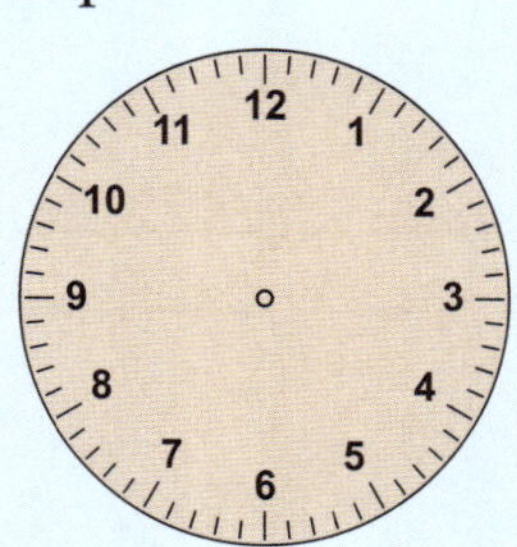

l quarter past 3

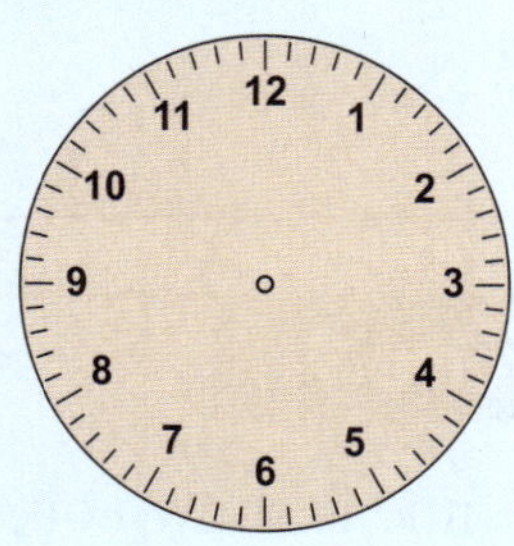

m half past 9

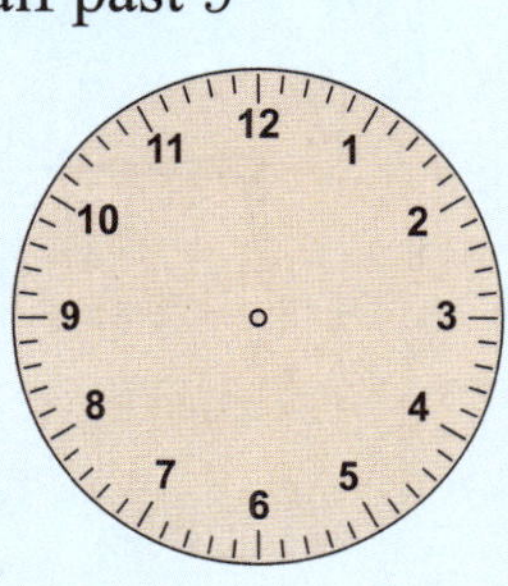

n quarter to 7

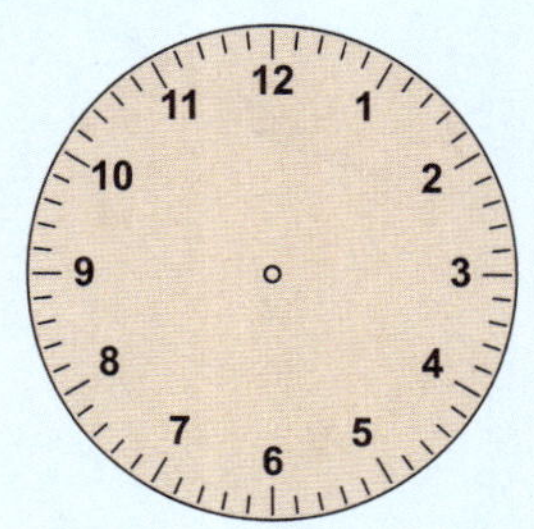

o quarter past 9

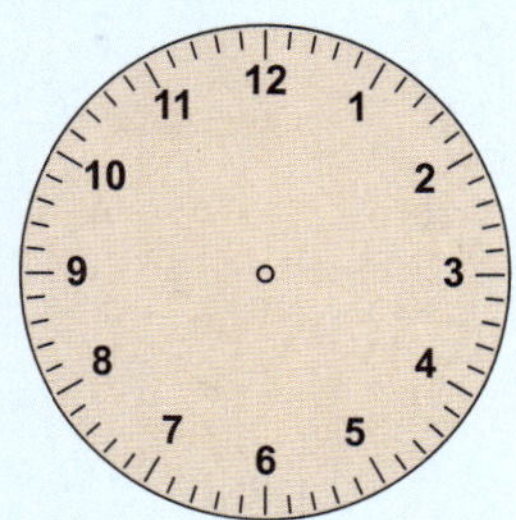

There are 60 **minutes** in each **hour**.

Each interval between the *numbers* on the clock is 5 minutes

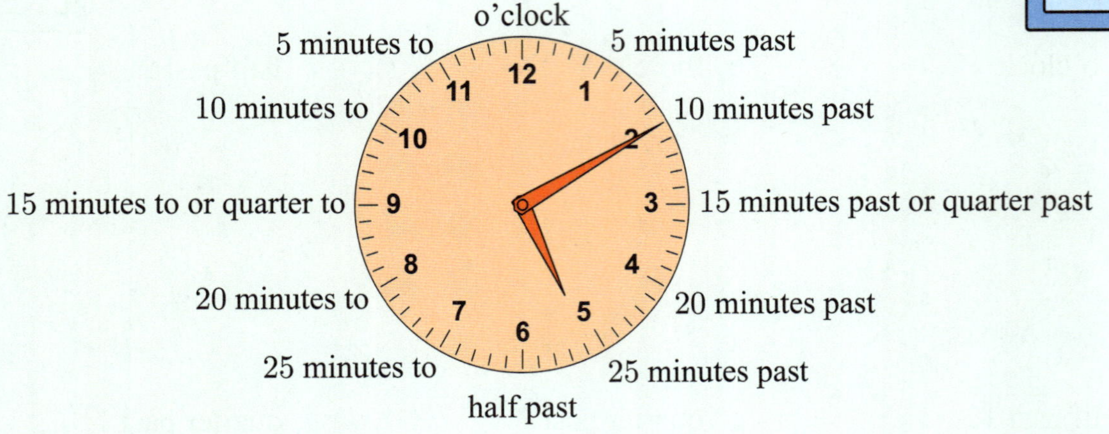

The minute hand shows 10 minutes past the hour.

The hour hand is between 5 and 6.

The time is "10 minutes past 5 o'clock" or "10 past 5".

Exercise 8

Write how many minutes it takes for the minute hand to move from 12 to each number of the clock.

These numbers are the results of the _______ times table.

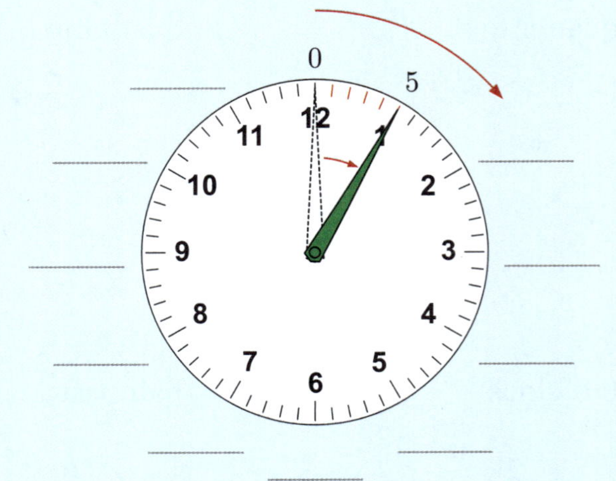

Exercise 9

What time is shown on the clock?

a

__________ past __________

b

____

c
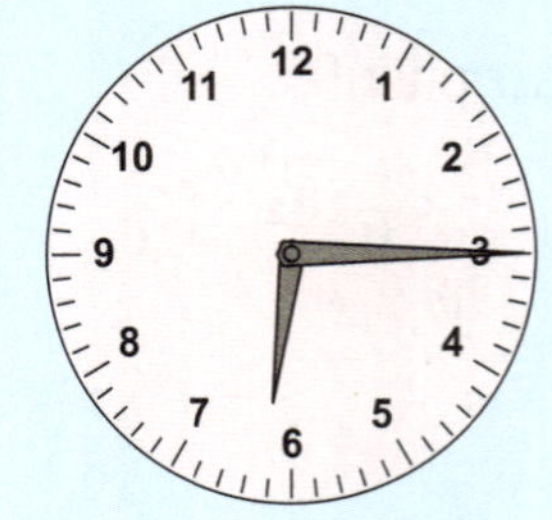

____

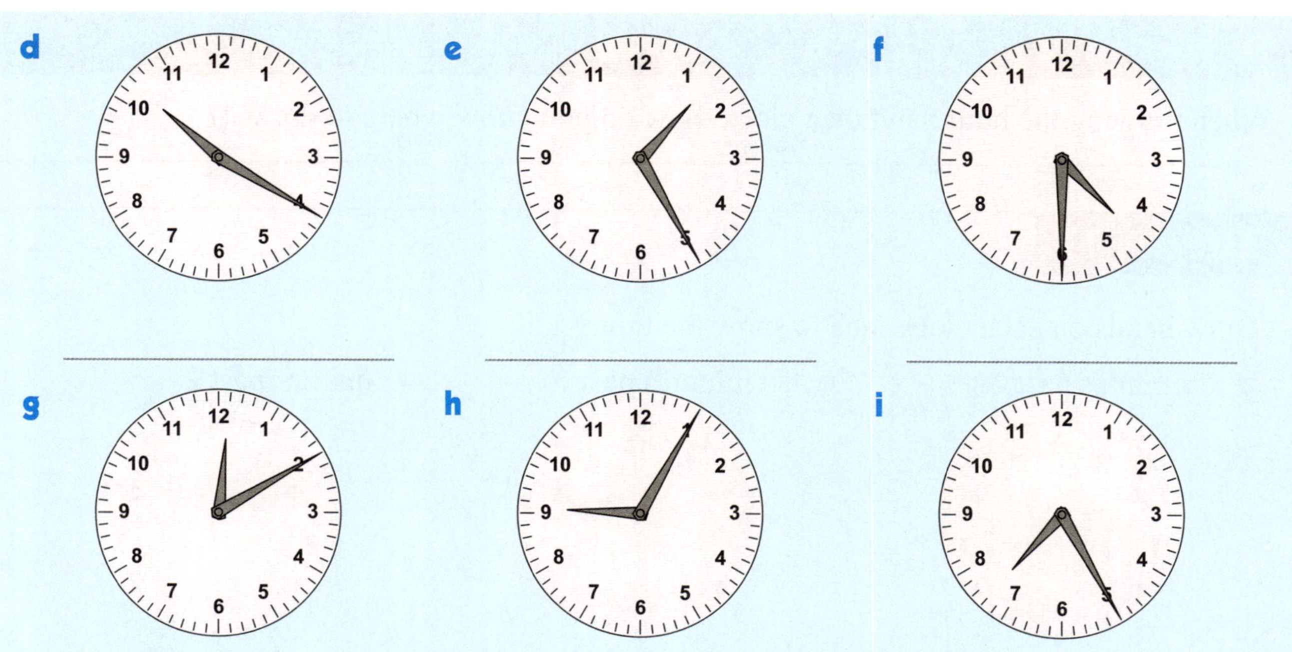

d   **e**   **f**

_______   _______   _______

g   **h**   **i**

_______   _______   _______

Exercise 10

What time is shown on the clock?

a   **b**   **c**

_______ to _______   _______   _______

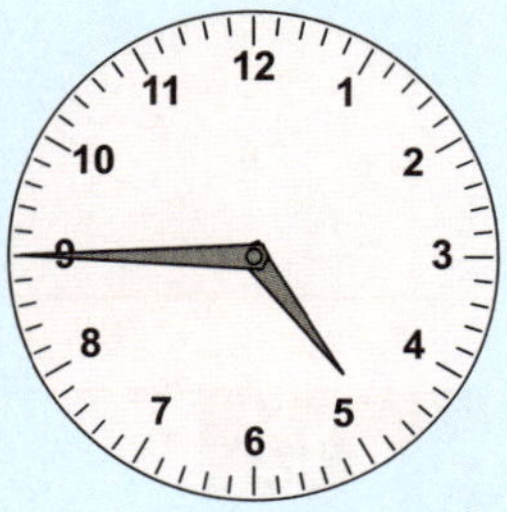

d    **e**    **f**

_______   _______   _______

g    **h**    **i**

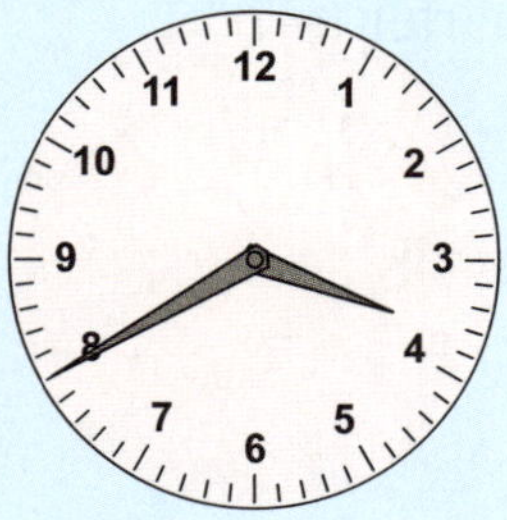

_______   _______   _______

Discussion

When drawing the hour hand on a clock, how do you know where to draw it?

Exercise 11

Draw hands on each clock face to show the time:

a 5 minutes past 3

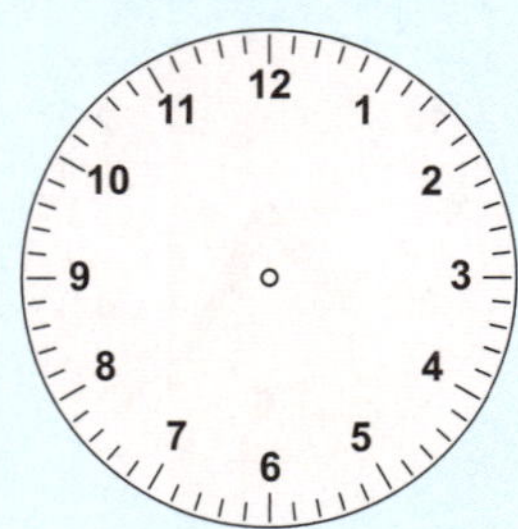

b 10 minutes past 6

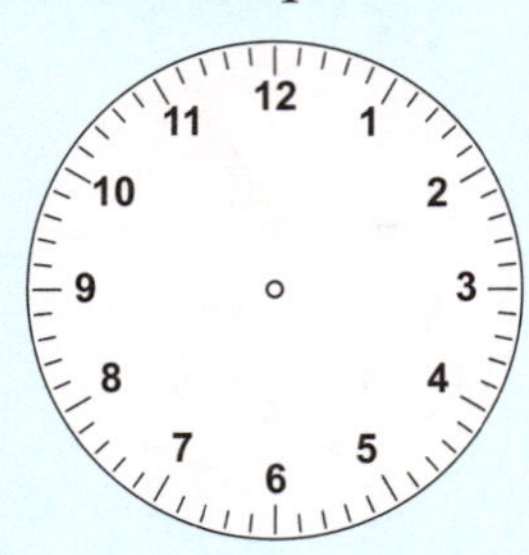

c quarter past 2

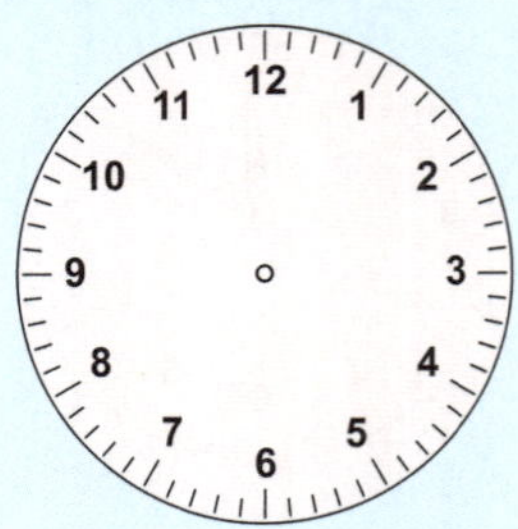

d 15 minutes past 4

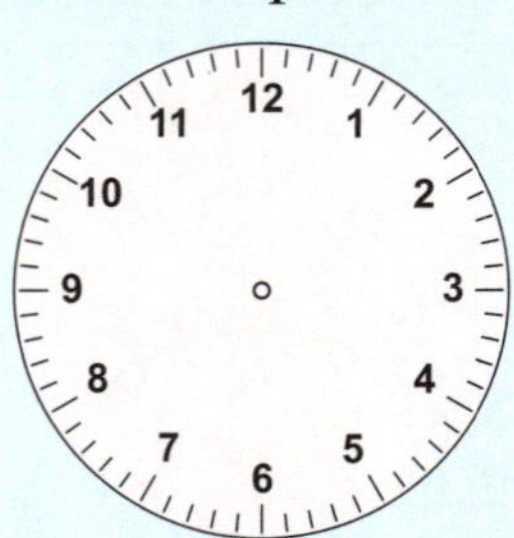

e 20 minutes past 7

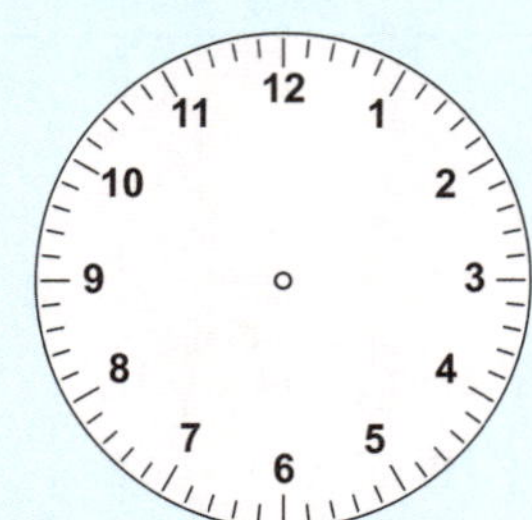

f 25 minutes past 6

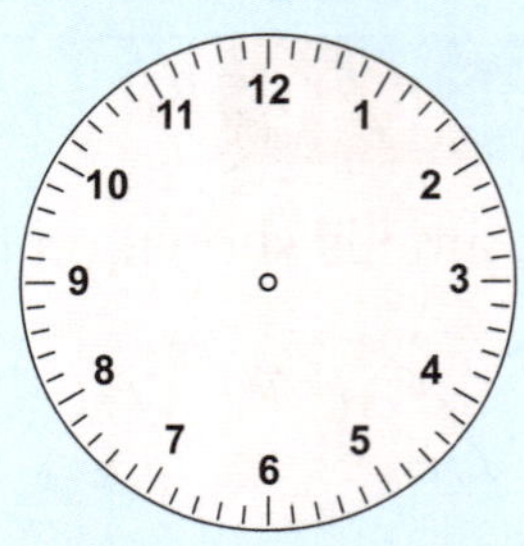

Exercise 12

Draw hands on each clock face to show the time:

a 25 minutes to 1

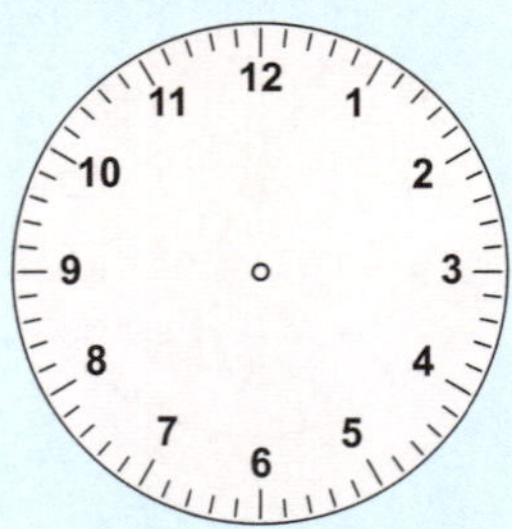

b 20 minutes to 8

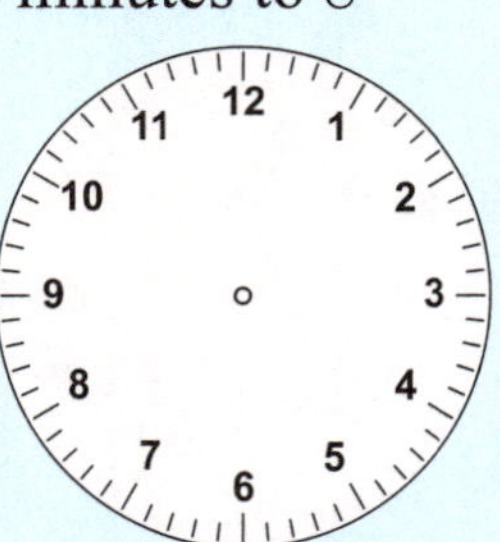

c 15 minutes to 5

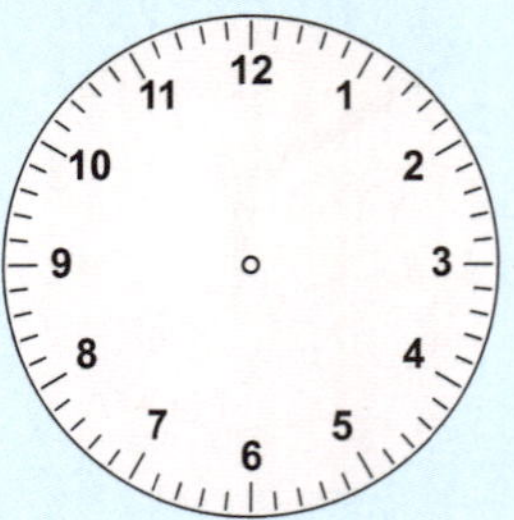

d quarter to 7

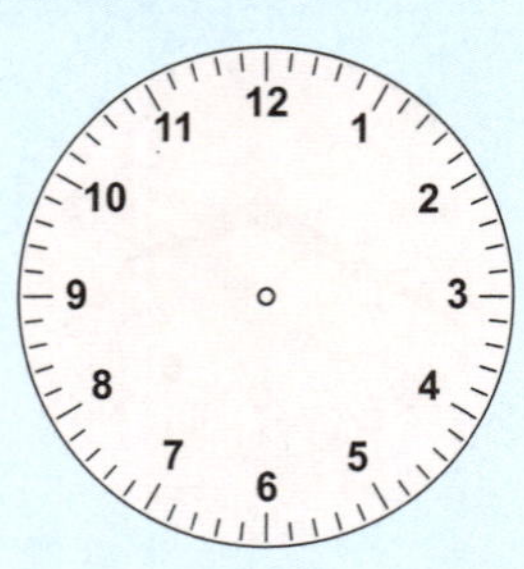

e 10 minutes to 11

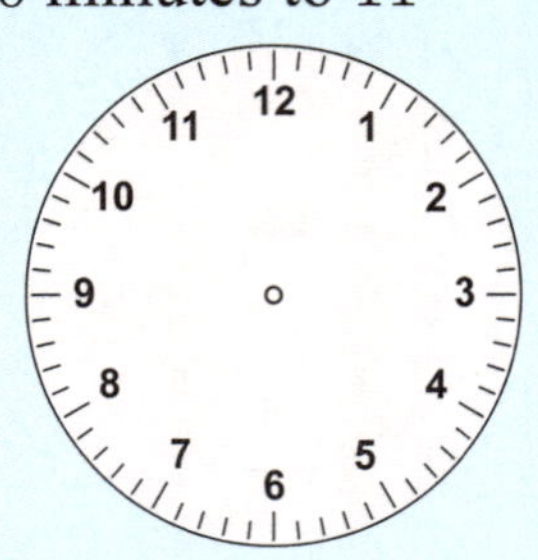

f 5 minutes to 9

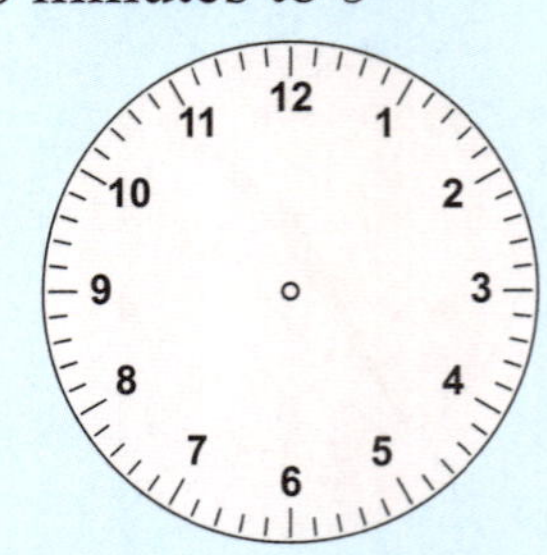

Exercise 13

Draw hands on each clock face to show the time:

a 10 past 9

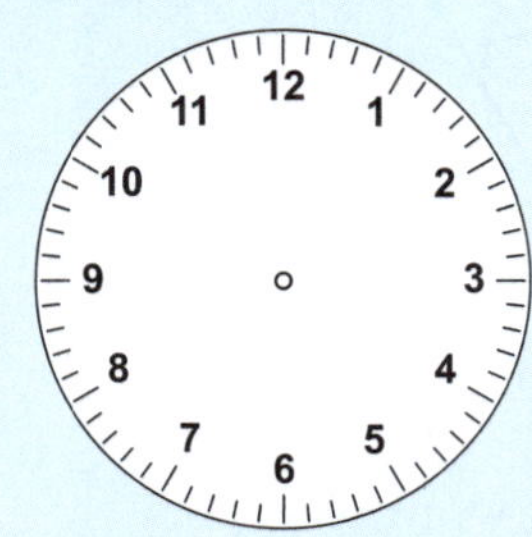

b 20 to 10

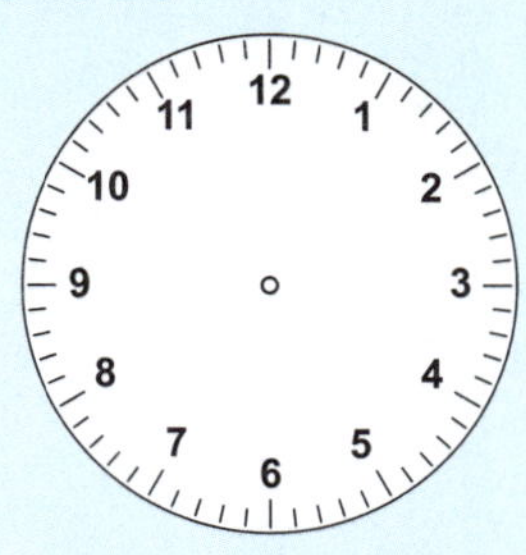

c 25 past 2

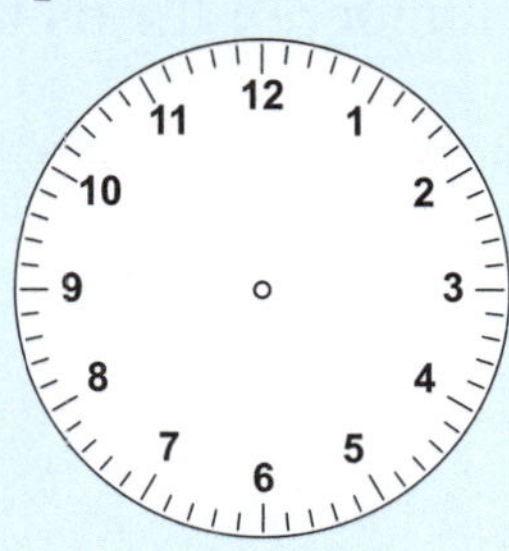

d 5 to 3

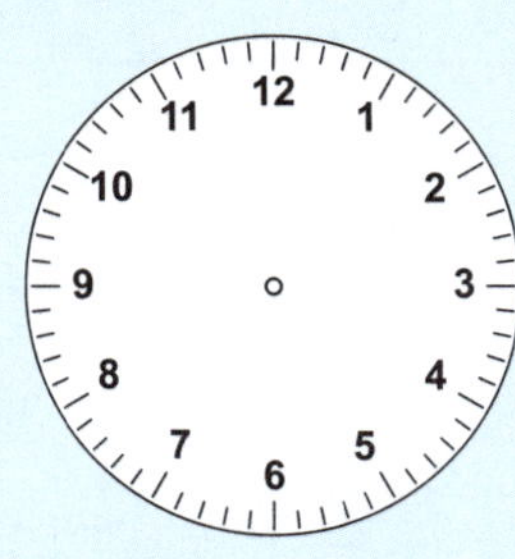

e 20 past 4

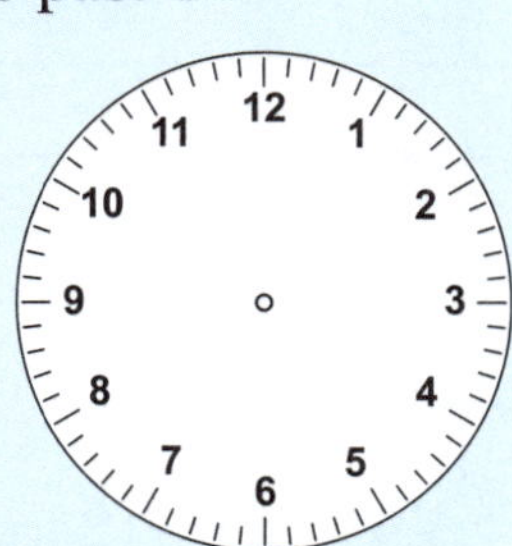

f 10 to 8

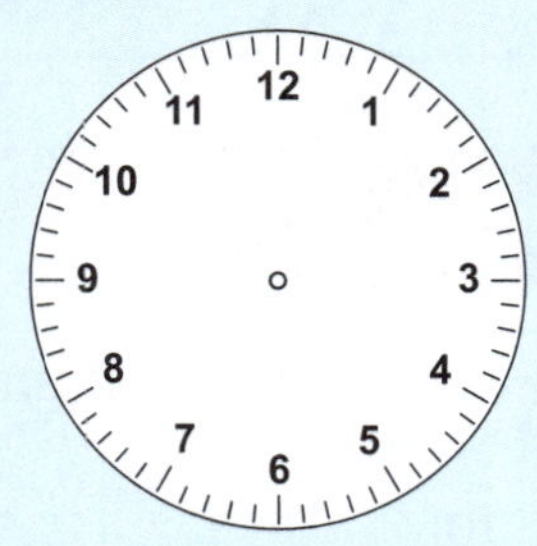

Exercise 14

Write how long it will take for the minute hand to move from:

a 12 to 6 _______________

b 12 to 4 _______________

c 1 to 3 _______________

d 2 to 7 _______________

e 4 to 6 _______________

f 6 to 9 _______________

g 7 to 8 _______________

h 6 to 12 _______________

i 8 to 11 _______________

j 9 to 12 _______________

k 11 to 1 _______________

l 8 to 3 _______________

There are 60 **seconds** in one minute.

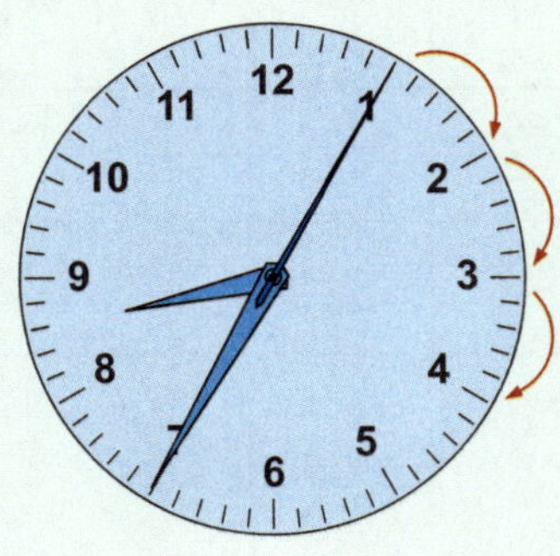

This clock has a **second hand**.

Each interval between the numbers on the clock is 5 seconds.

The second hand will take 15 seconds to count on from 1 to 4.

Exercise 15

Write how many seconds it takes for the second hand to move from 12 to each number of the clock.

These numbers are the results of the ________ times table.

Exercise 16

Write how long it will take for the second hand to move from:

a 12 to 3

b 12 to 7

c 3 to 5

d 5 to 9

e 6 to 7

f 8 to 11

g 11 to 4

h 1 to 9

i 3 to 12

j 10 to 5

Activity

If you are asked to count seconds, it is very easy to count too fast or too slow.

You can use words to space out the numbers.

For example: "1 potato,   2 potato,   3 potato,   4 potato, ..."

or   "1 elephant,   2 elephant,   3 elephant,   4 elephant, ..."

Your teacher will use a stopwatch to see how well you can guess 60 seconds. Record the number you have counted to when your teacher says "stop".

Alternatively, click the icon to practise your time estimation skills.

Trial	Your count
1	
2	
3	

Exercise 17

Complete:

There are _______ days in a **week**.

In order, they are

S u ________________

M ________________

T ________________

W ________________

T h ________________

F ________________

S a ________________

The days of the **weekend** are ________________ and ________________ .

Exercise 18

Complete:

There are _______ **months** in a **year**.

In order, they are

J ________________

F ________________

M ________________

A ________________

M ________________

J ________________

J ________________

A ________________

S ________________

O ________________

N ________________

D ________________

Discussion

How do we number the years?

Why were they numbered this way?

A **calendar** shows the months and days of the year.

Some months have more days than others.

Calendar 2027

January

Su	M	T	W	Th	F	Sa
					1	2
3	4	5	6	7	8	9
10	11	12	13	14	15	16
17	18	19	20	21	22	23
24	25	26	27	28	29	30
31						

February

Su	M	T	W	Th	F	Sa
	1	2	3	4	5	6
7	8	9	10	11	12	13
14	15	16	17	18	19	20
21	22	23	24	25	26	27
28						

March

Su	M	T	W	Th	F	Sa
	1	2	3	4	5	6
7	8	9	10	11	12	13
14	15	16	17	18	19	20
21	22	23	24	25	26	27
28	29	30	31			

April

Su	M	T	W	Th	F	Sa
				1	2	3
4	5	6	7	8	9	10
11	12	13	14	15	16	17
18	19	20	21	22	23	24
25	26	27	28	29	30	

May

Su	M	T	W	Th	F	Sa
						1
2	3	4	5	6	7	8
9	10	11	12	13	14	15
16	17	18	19	20	21	22
23	24	25	26	27	28	29
30	31					

June

Su	M	T	W	Th	F	Sa
		1	2	3	4	5
6	7	8	9	10	11	12
13	14	15	16	17	18	19
20	21	22	23	24	25	26
27	28	29	30			

July

Su	M	T	W	Th	F	Sa
				1	2	3
4	5	6	7	8	9	10
11	12	13	14	15	16	17
18	19	20	21	22	23	24
25	26	27	28	29	30	31

August

Su	M	T	W	Th	F	Sa
1	2	3	4	5	6	7
8	9	10	11	12	13	14
15	16	17	18	19	20	21
22	23	24	25	26	27	28
29	30	31				

September

Su	M	T	W	Th	F	Sa
		1	2	3	4	
5	6	7	8	9	10	11
12	13	14	15	16	17	18
19	20	21	22	23	24	25
26	27	28	29	30		

October

Su	M	T	W	Th	F	Sa
					1	2
3	4	5	6	7	8	9
10	11	12	13	14	15	16
17	18	19	20	21	22	23
24	25	26	27	28	29	30
31						

November

Su	M	T	W	Th	F	Sa
	1	2	3	4	5	6
7	8	9	10	11	12	13
14	15	16	17	18	19	20
21	22	23	24	25	26	27
28	29	30				

December

Su	M	T	W	Th	F	Sa
		1	2	3	4	
5	6	7	8	9	10	11
12	13	14	15	16	17	18
19	20	21	22	23	24	25
26	27	28	29	30	31	

Exercise 19

Write down the number of days in each month in a *normal* year.

Month	Days
January	
February	
March	
April	
May	
June	

Month	Days
July	
August	
September	
October	
November	
December	

Discussion

What is a **leap year**? Why do we have leap years?

How many days does February have in a leap year? ________________

How often do we usually have leap years? ________________________

When is the *next* leap year? ________________

Exercise 20

July 2027						
Su	M	T	W	Th	F	Sa
				1	2	3
4	5	6	7	8	9	10
11	12	13	14	(15)	16	17
18	19	20	21	22	23	24
25	26	27	28	29	30	31

This calendar shows the days of the week in the month of July, 2027.

The **date** of the day circled in red is Thursday, 15th July.

We say "Thursday, the fifteenth of July".

a Use the calendar to fill in the day of the week:

________________________ , 5th July

________________________ , 24th July

________________________ , 13th July

________________________ , 11th July

b There are ________ Fridays in July 2027.

c The day after Tuesday, 6th July is

____ .

d The day before Monday, 19th July is

____ .

e The day after Saturday, 31st July is

____ .

f The day before Thursday, 1st July is

____ .

Exercise 21

Write the date of the day *after*:

a   Tuesday, 19th August

b   Friday, 6th February

c   Sunday, 30th April

Exercise 22

Write the date of the day *before*:

a   Friday, 11th May

b   Monday, 28th December

c   Thursday, 1st November

Exercise 23

December 2027						
Su	M	T	W	Th	F	Sa
			1	2	3	4
5	6	7	8	9	10	11
12	13	14	15	16	17	18
19	20	21	22	23	24	25
26	27	28	29	30	31	

This calendar shows the days of the week in the month of December, 2027.

a   Write the date of the day *two days after* Thursday, 16th December.

b   Write the date of the day *three days before* Monday, 27th December.

c   Write the date of the day *one week after* Wednesday, 8th December.

d   Write the date of the day *two weeks before* Sunday, 19th December.

Discussion

Depending on your location, you might experience **seasons**.

Click the icon to find out *why* we have seasons.

I live in the ________________________ hemisphere.

In this hemisphere, the seasons are:

Summer	**Autumn or Fall**
___________________ ___________________ ___________________	___________________ ___________________ ___________________
Winter	**Spring**
___________________ ___________________ ___________________	___________________ ___________________ ___________________

Revision

1 Complete using *now*, *past*, and *future*:

Yesterday is in the ________________ .

Tomorrow is in the ________________ .

Today is the day right ________________ .

2 Complete:

Each day is divided into _______ hours.

There are _______ hours from ___________________ until noon.

These are am times which make up the ___________________ .

There are _______ hours from noon until ___________________ .

These are _______ times which make up the afternoon and evening.

3 What time is shown?

a 

b 

c 

d 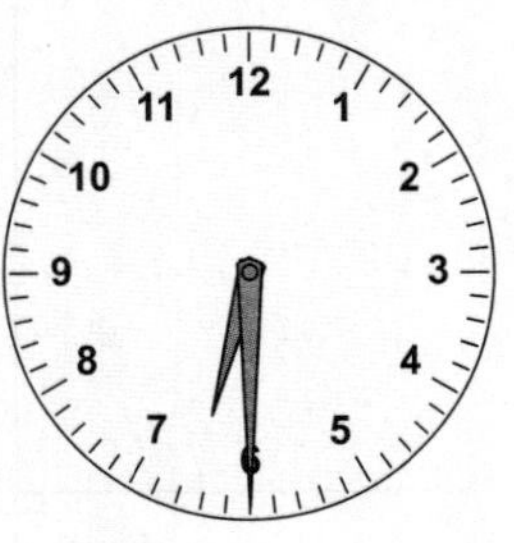

e 

f 

4 Draw hands on each clock to show the time:

a quarter to 5

b 10 o'clock

c half past 8

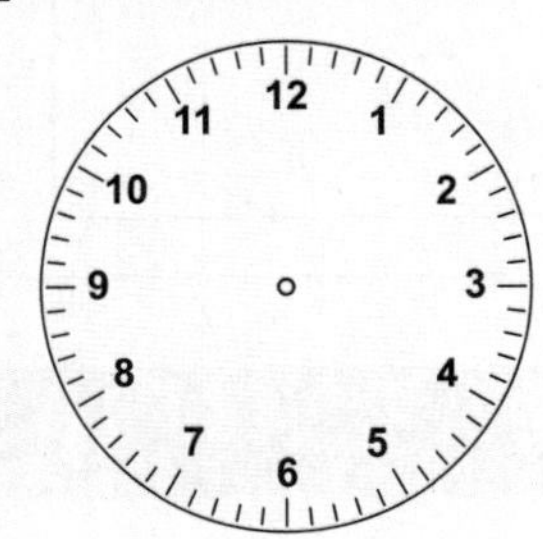

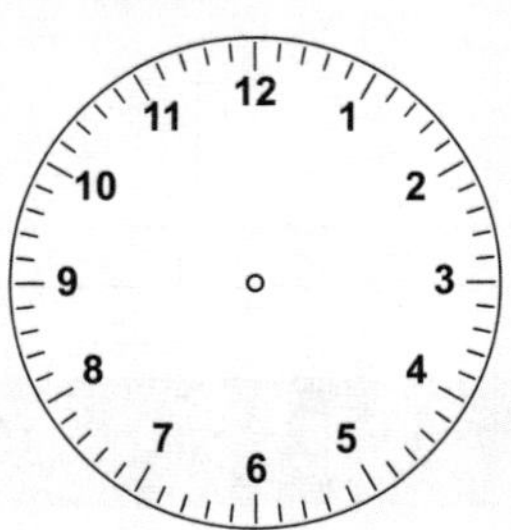

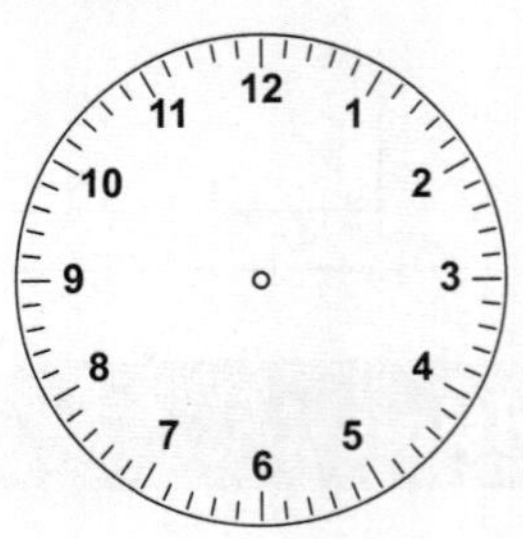

d quarter past 2

e half past 11

f 6 o'clock

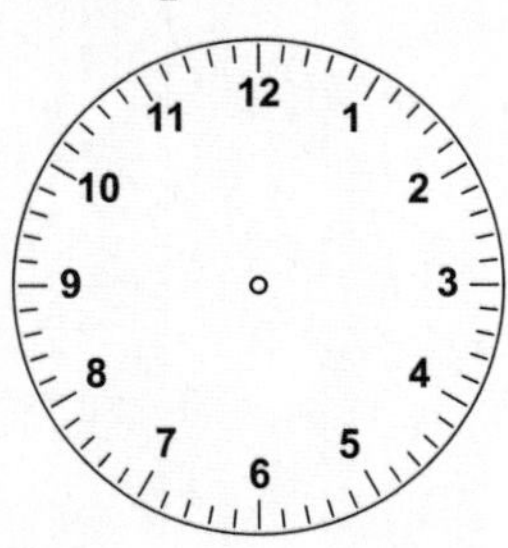

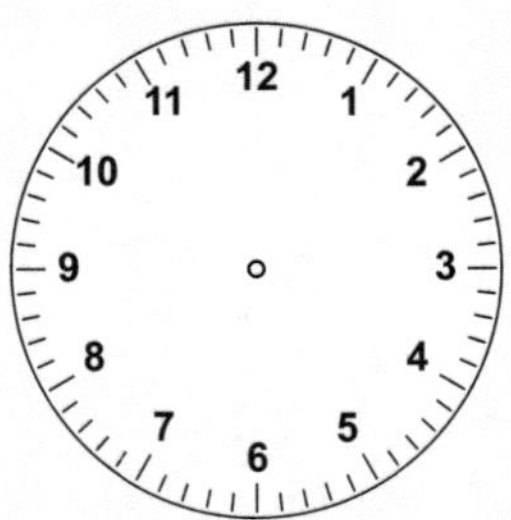

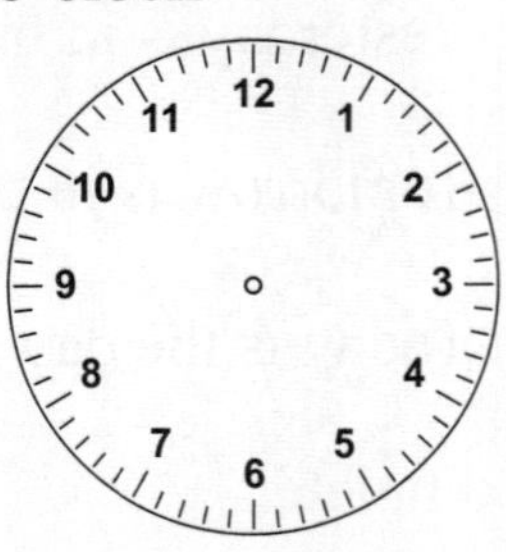

g quarter past 12

h quarter to 3

i half past 4

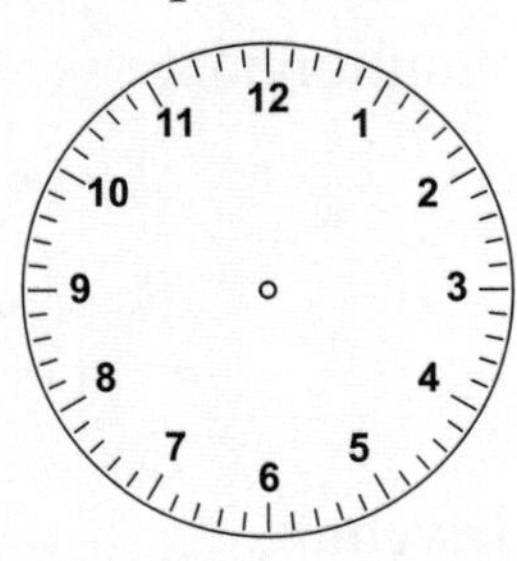

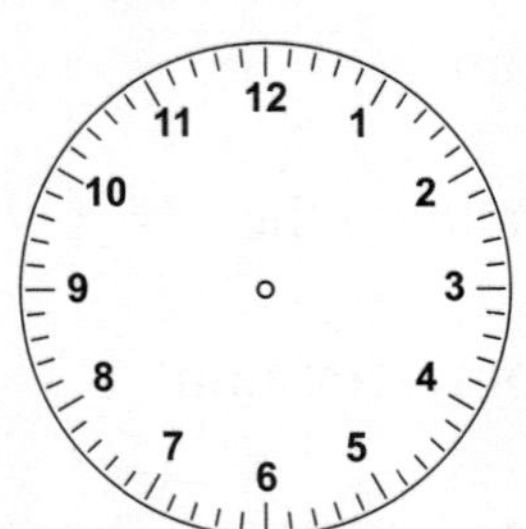

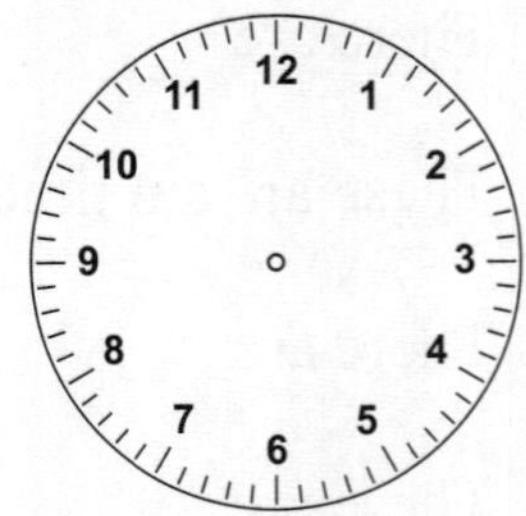

5 What time is shown?

a 

b 

c 

d 

e 

f 

6 Draw hands on each clock to show the time:

a 25 minutes past 3

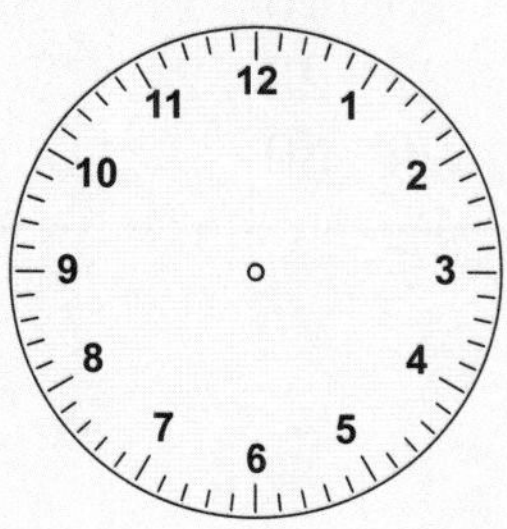

b 10 minutes to 7

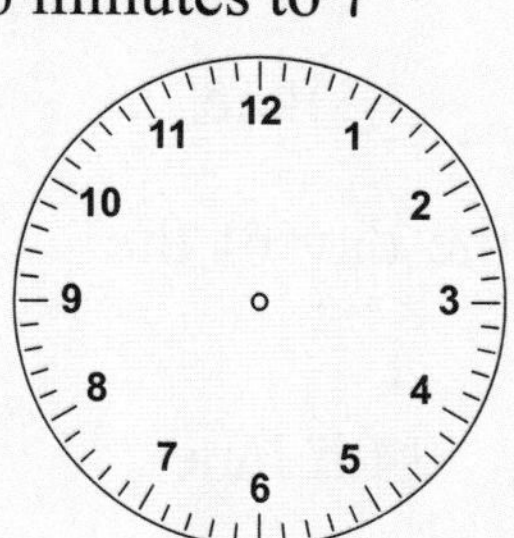

c 5 minutes past 10

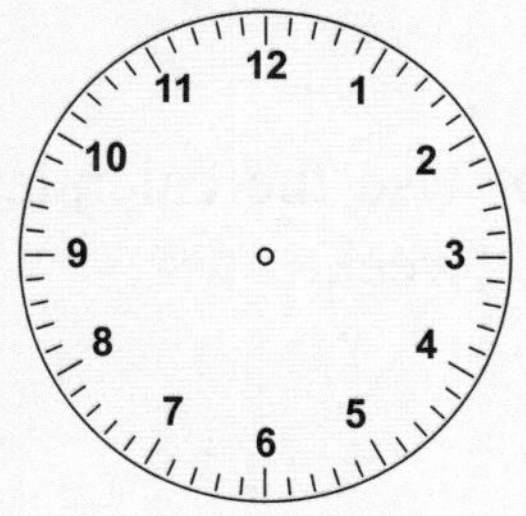

d 25 minutes to 6

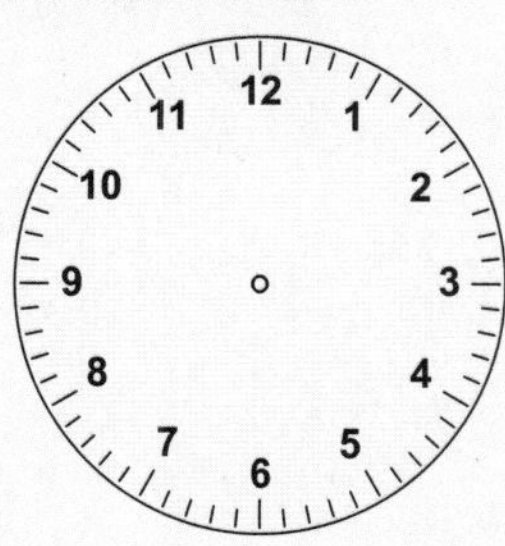

e 20 minutes to 12

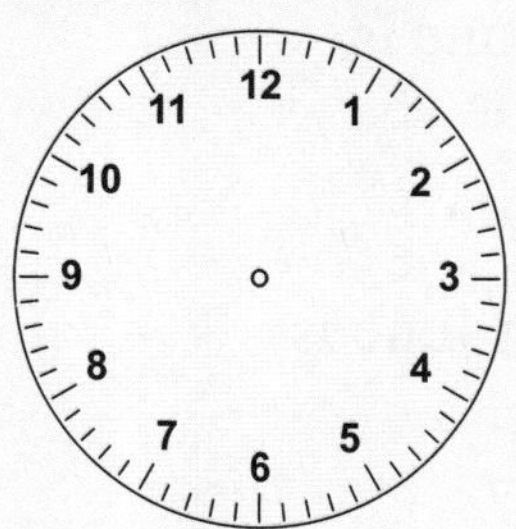

f 10 minutes past 11

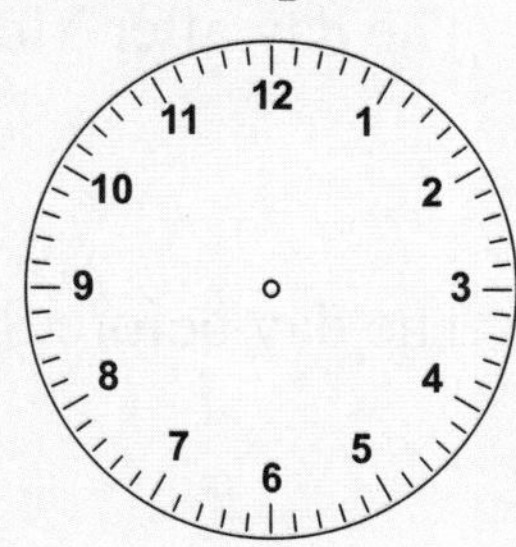

g 5 to 9

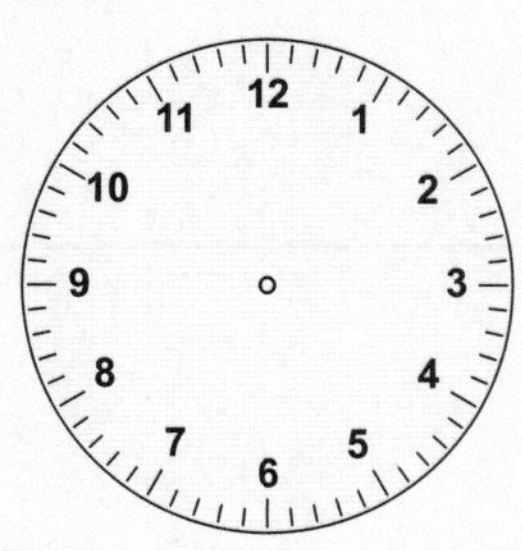

h 20 past 8

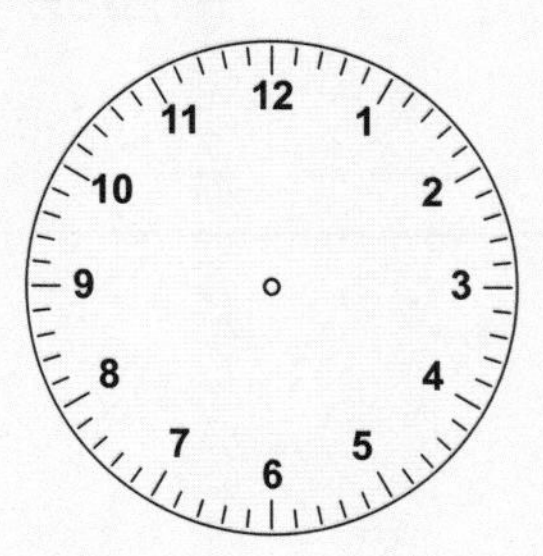

i 10 to 10

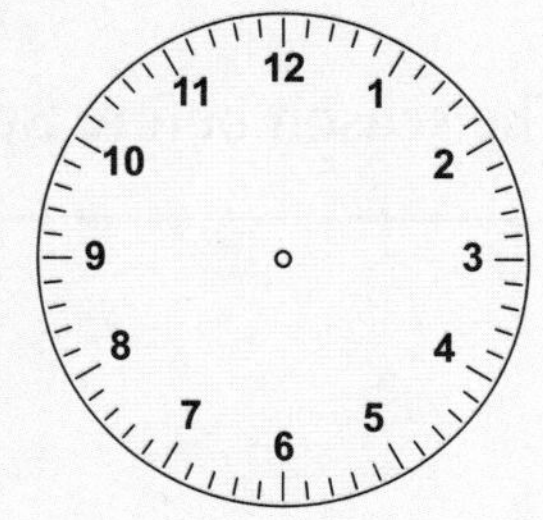

7 Write how long it will take for the minute hand to move from:

a 12 to 8 _______________   **b** 3 to 7 _______________

c 5 to 10 _______________   **d** 9 to 1 _______________

8 Write how long it will take for the second hand to move from:

a 12 to 6 _______________   **b** 4 to 7 _______________

c 3 to 11 _______________   **d** 7 to 2 _______________

9 Write down the number of days in each month:

a March _______________   **b** September _______________

c December _______________   **d** June _______________

10 This calendar shows the days of the week in June 2027.

a Julie's birthday is marked with the red circle. Her birthday is

_______________ , _______ June.

b Use the calendar to fill in the day of the week:

_______________ , 14th June

_______________ , 3rd June

c The day after Monday, 7th June is

_______________________________ .

d The day before Sunday, 27th June is

_______________________________ .

June 2027						
Su	M	T	W	Th	F	Sa
		1	2	3	4	5
6	7	8	9	10	11	12
13	14	15	16	17	18	19
20	21	22	23	24	(25)	26
27	28	29	30			

11 The season after Summer is _______________ .

The season before Spring is _______________ .

Wendy's ceiling fan is controlled by a dial.

0 is off. There are 3 speeds.

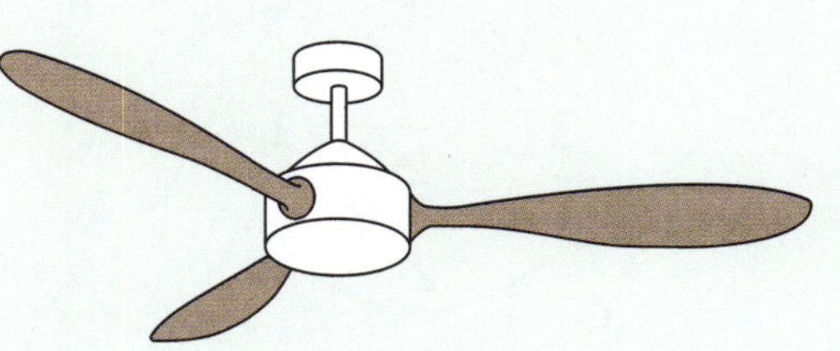

- If Wendy turns the dial all the way around, it is a **full turn**.

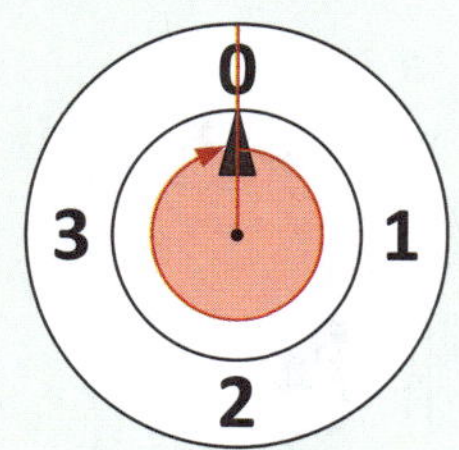

- If Wendy turns the dial from 0 to 2, it is a **half turn**.

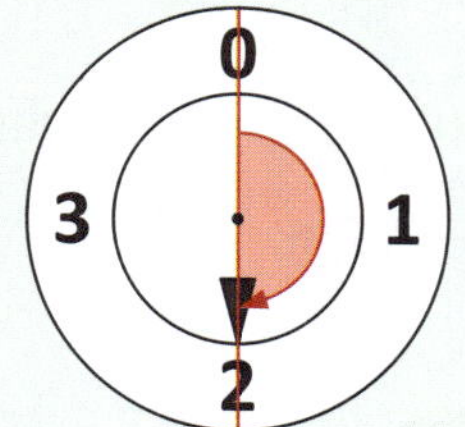

- If Wendy turns the dial from 2 to 3, it is a **quarter turn**.

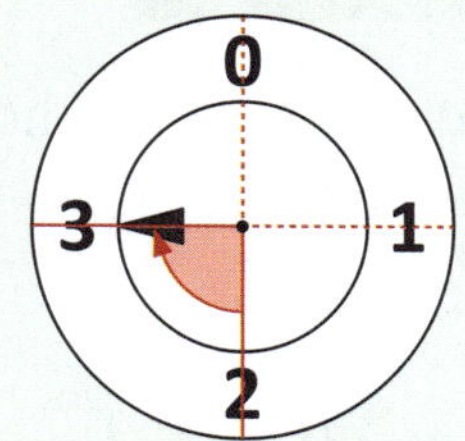

Exercise 1

In one hour, the minute hand makes a

____

turn.

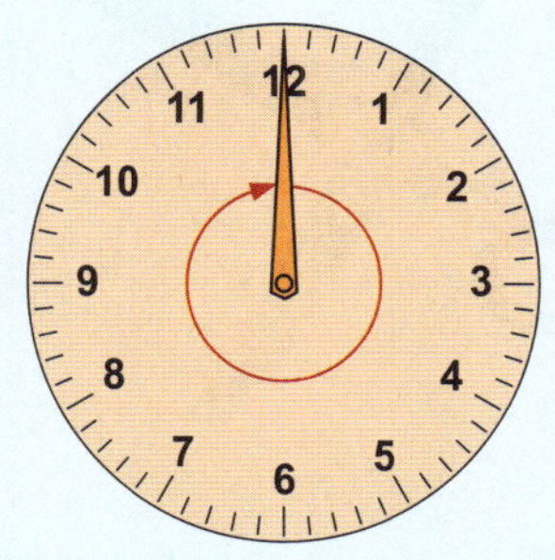

In half an hour, the minute hand makes a

____

turn.

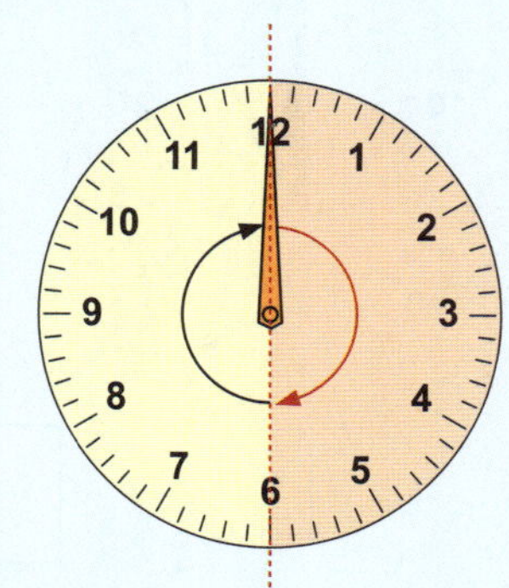

In a quarter of an hour, the minute hand makes a

____

turn.

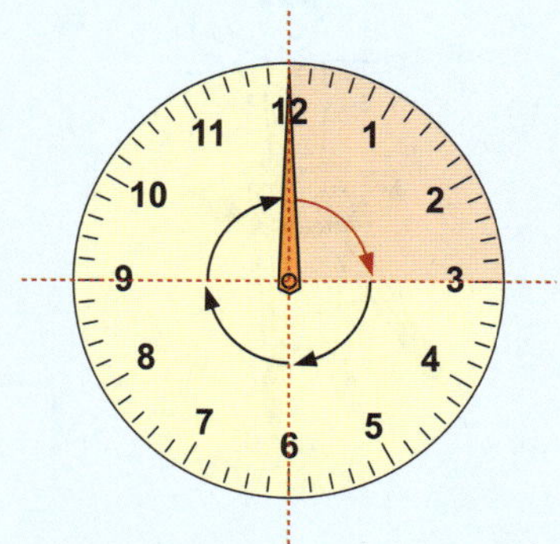

Exercise 2

Describe each turn:

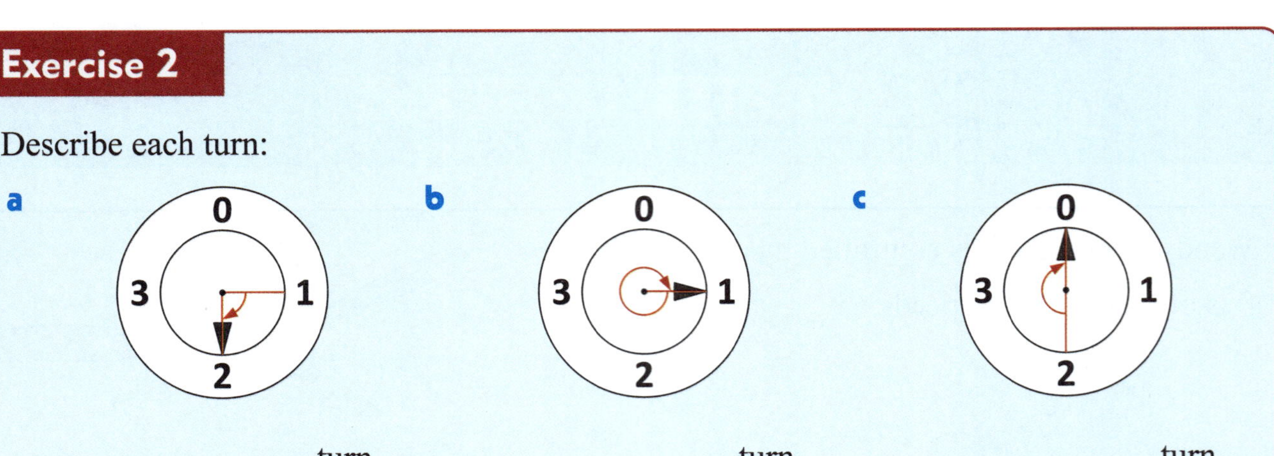

a ______________________ turn

b ______________________ turn

c ______________________ turn

Exercise 3

Place a tick ✓ in the box if the picture shows a quarter turn. Place a cross ✗ in the box if it does not.

a □

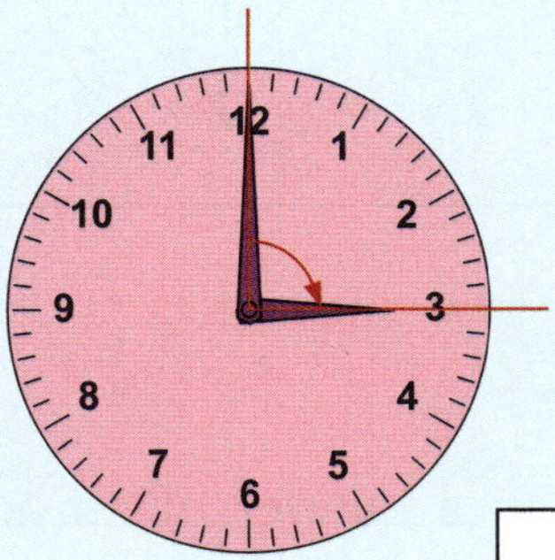

b □

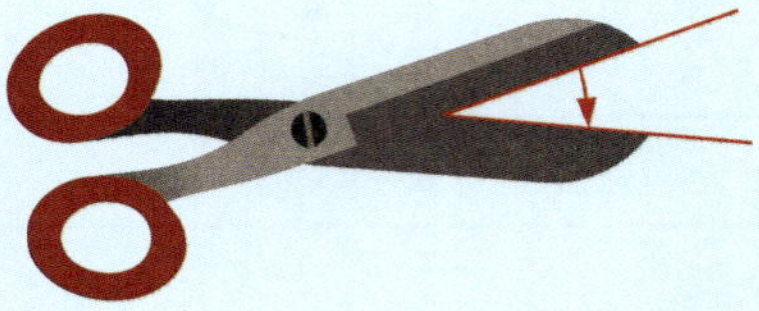

c □

d □

e □

f □

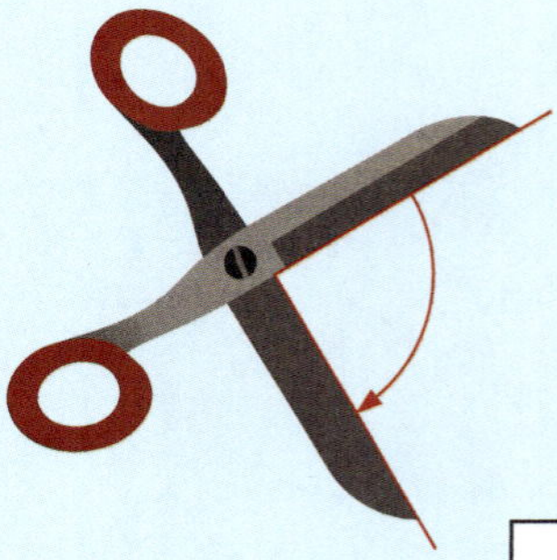

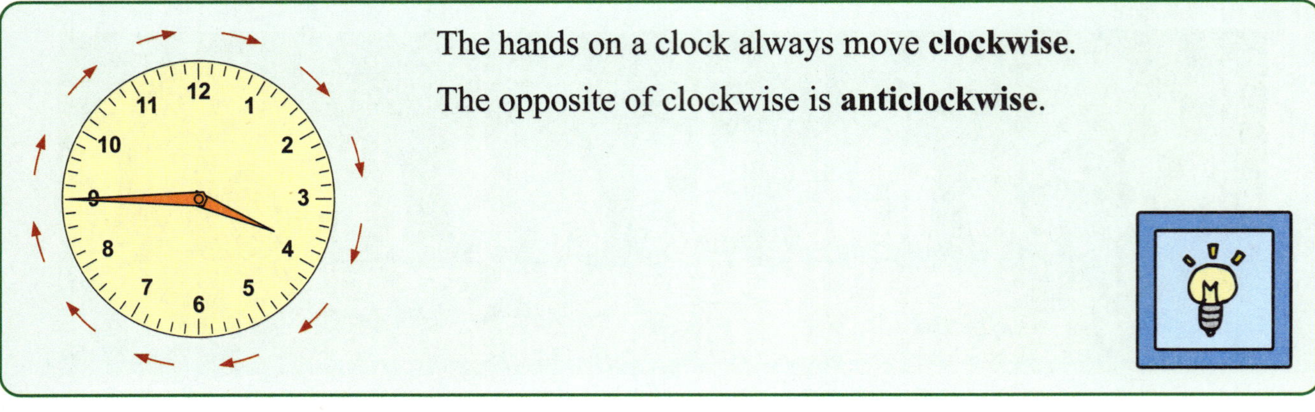

The hands on a clock always move **clockwise**.

The opposite of clockwise is **anticlockwise**.

Exercise 4

Decide whether each turn is clockwise or anticlockwise:

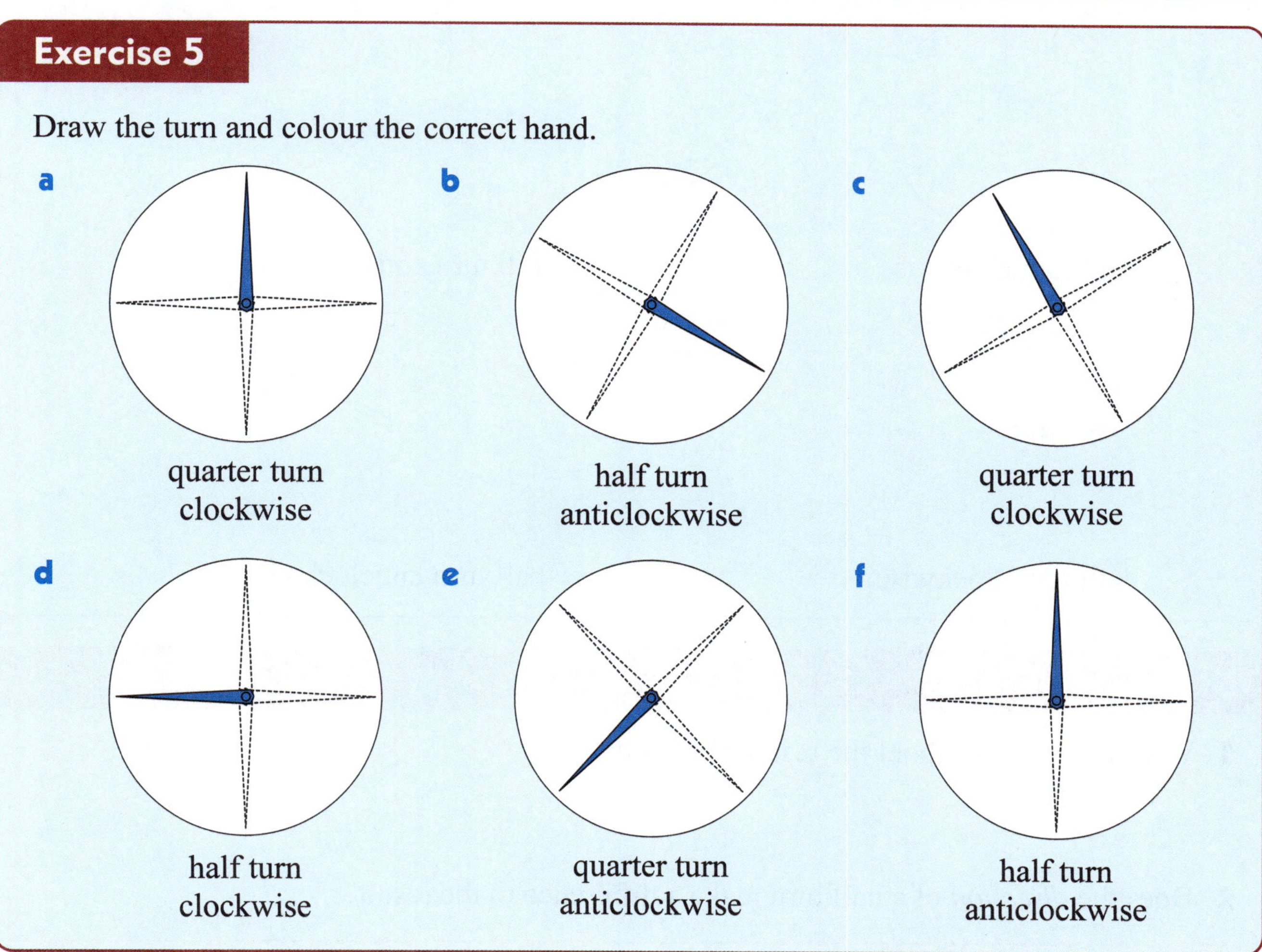

Exercise 5

Draw the turn and colour the correct hand.

a
quarter turn
clockwise

b
half turn
anticlockwise

c
quarter turn
clockwise

d
half turn
clockwise

e
quarter turn
anticlockwise

f
half turn
anticlockwise

Exercise 6

Draw the needle after a quarter turn in the given direction.

a
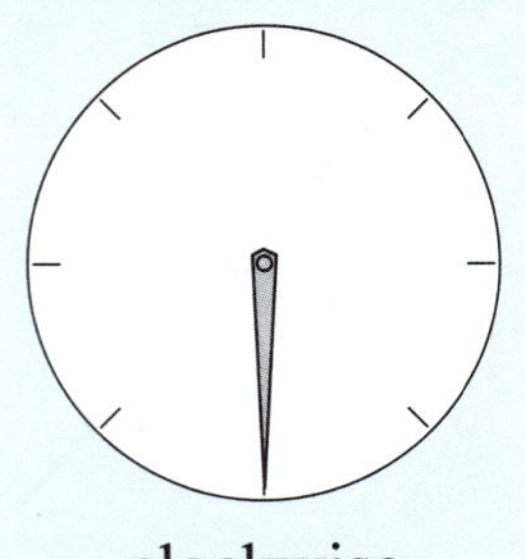
clockwise

b
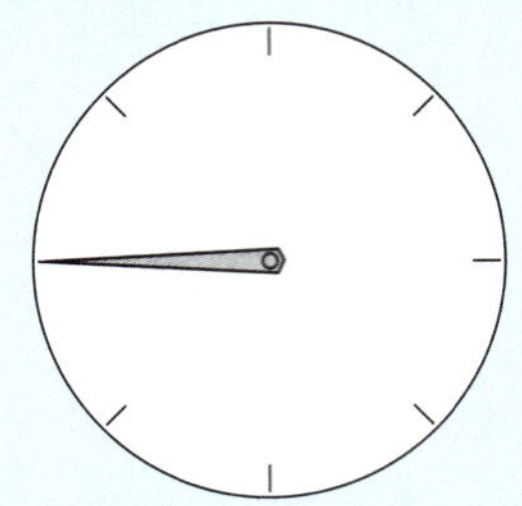
anticlockwise

c

clockwise

d
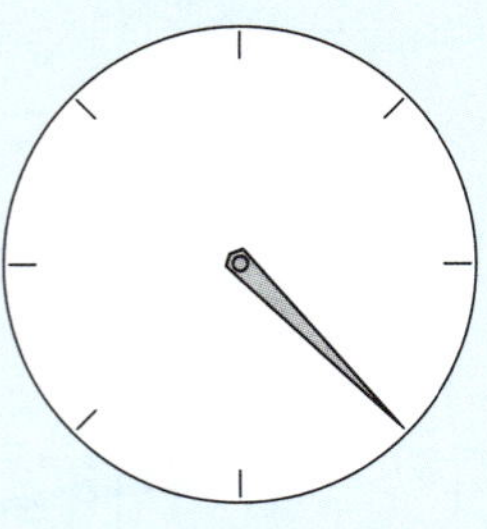
anticlockwise

e

clockwise

f

anticlockwise

Exercise 7

Draw a turn which is a:

a
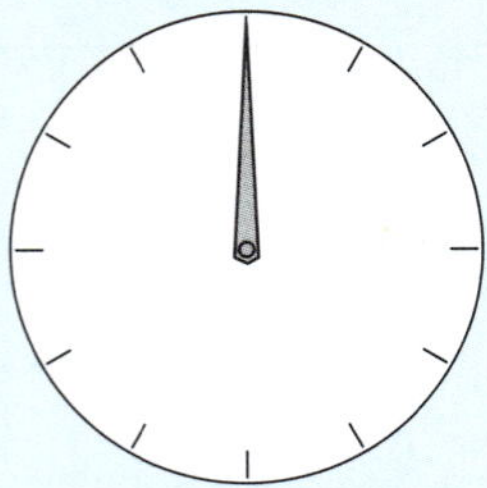
full turn clockwise

b

full turn anticlockwise

c
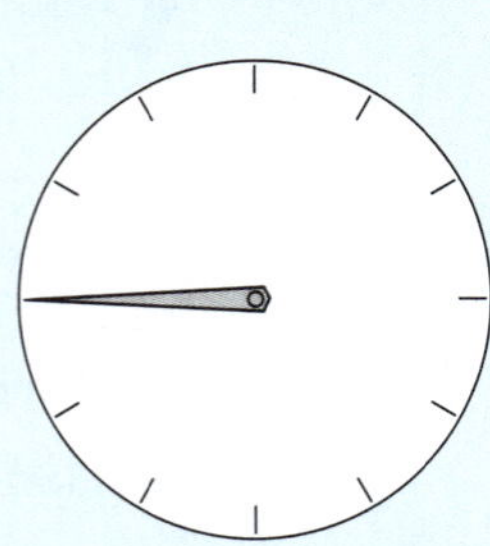
half turn clockwise

d

half turn anticlockwise

Discussion

1 What can you say about the result of a full turn?

____

2 Does the *direction* of a half turn make a difference to the result? _______________________

Game

Click on the icon to practise working with turns.

Activity   In your environment

You will need: pencils

What to do:

1   Move your seat to a different position in the classroom or take it outside.

2   Sit on your seat and face forwards.

3   Sketch something that you see straight in front of you into box 1.

4   Turn your seat a quarter turn to the right. Draw an object or person straight in front of you in box 2.

5   Turn another quarter turn and complete box 3.

6   Turn another quarter turn and complete box 4.

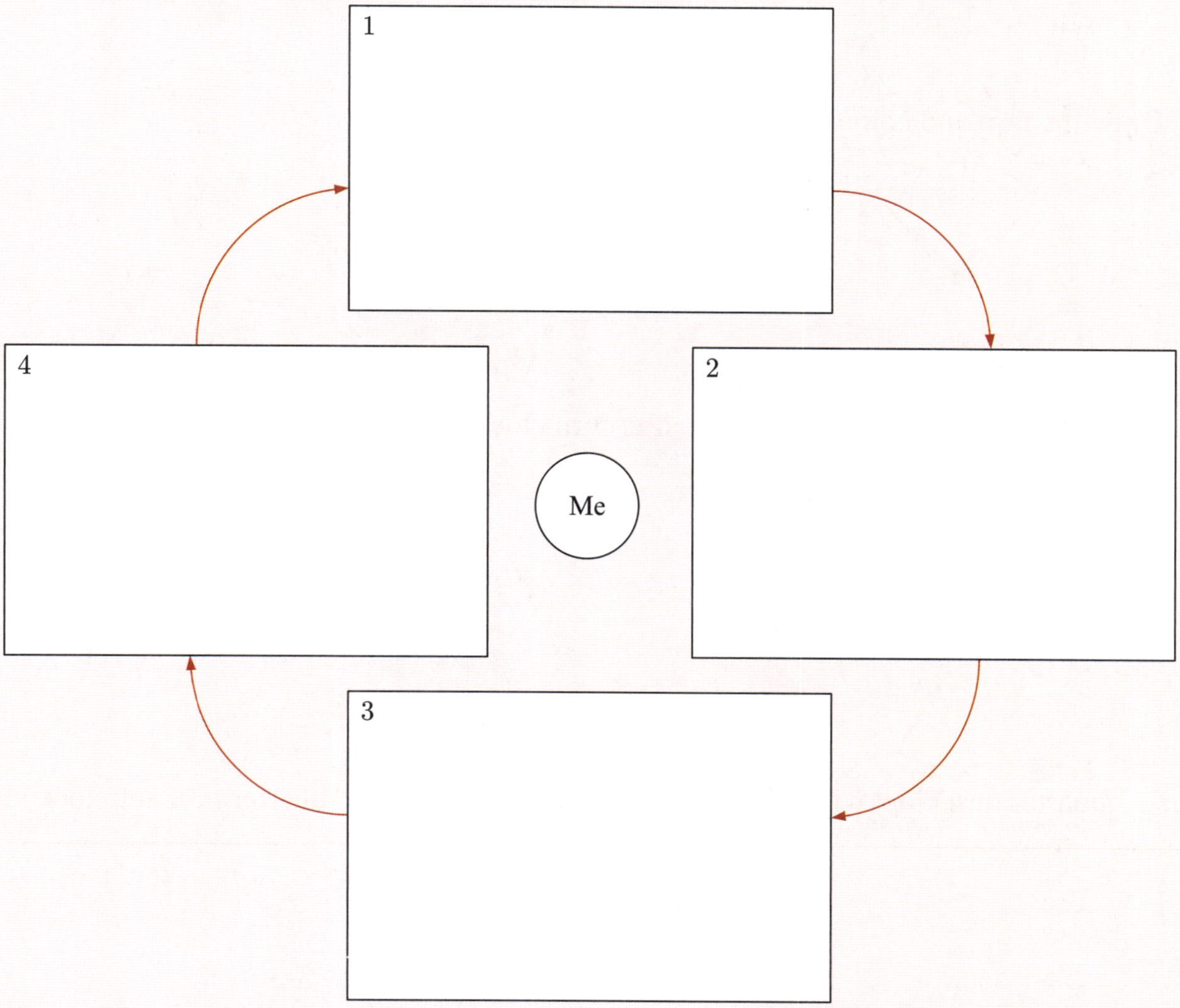

Revision

1 Describe each turn as a *quarter turn*, *half turn*, or *full turn*:

a 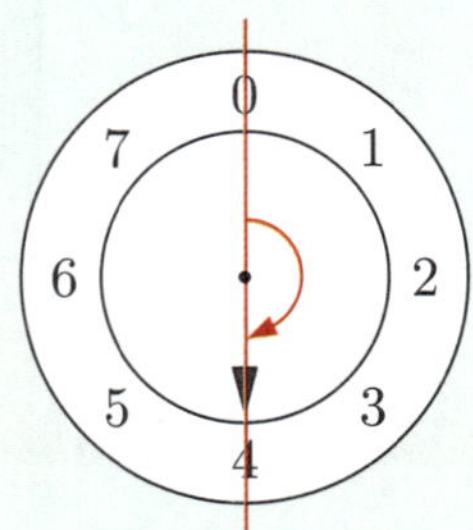

b 

c 

_______________ turn   _______________ turn   _______________ turn

2 Decide whether each turn is clockwise or anticlockwise:

a 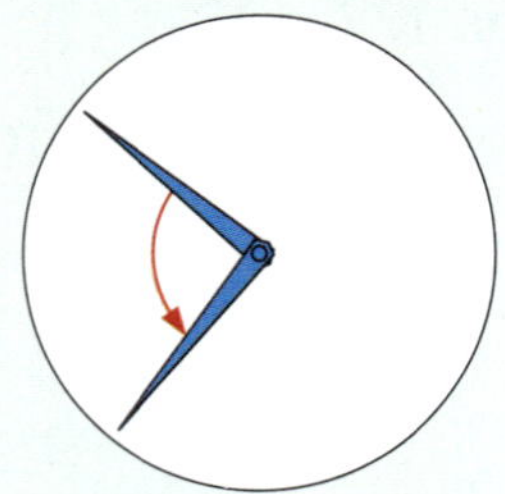

b 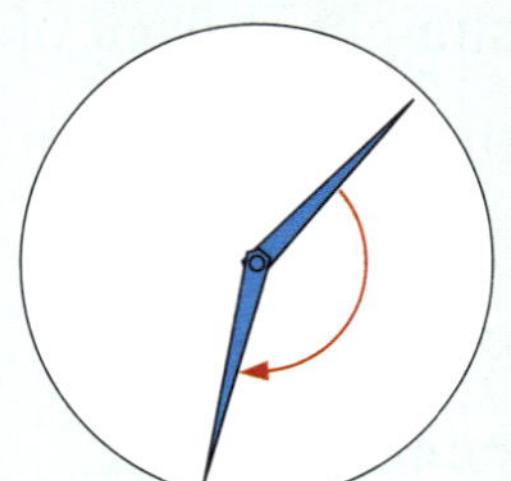

c 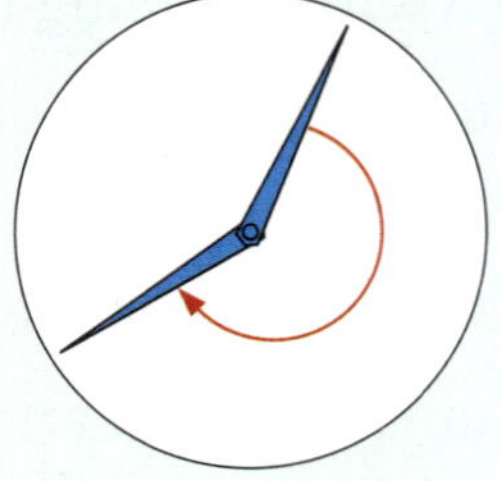

_______________   _______________   _______________

3 Draw the turn and colour the correct hand.

a 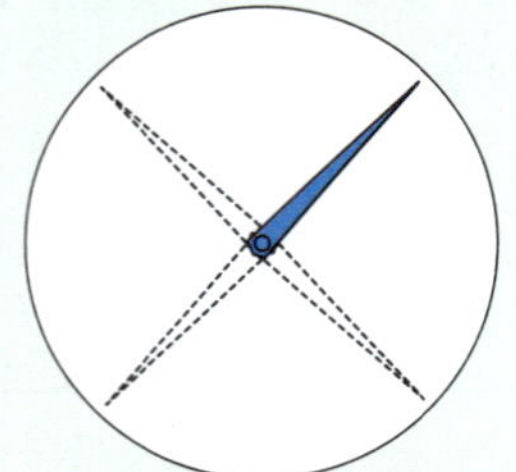

b 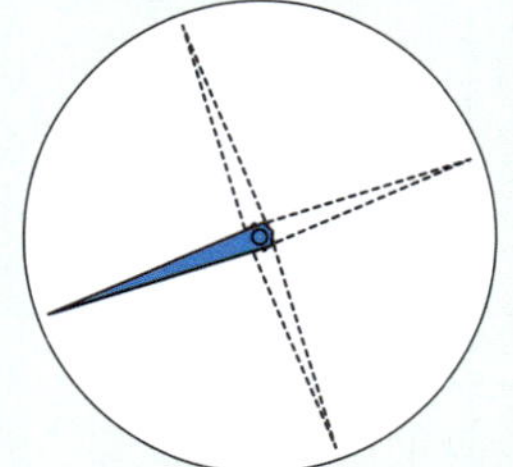

c 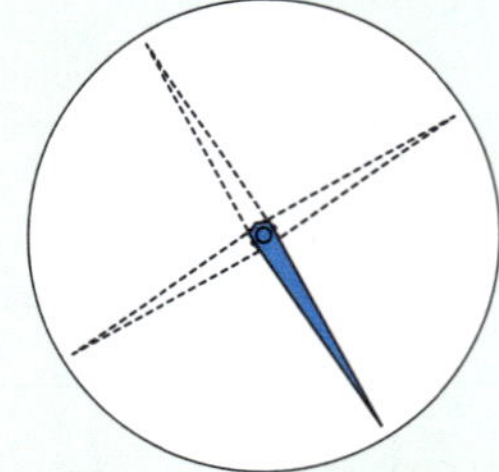

half turn anticlockwise   quarter turn clockwise   quarter turn anticlockwise

4 Draw the needle after the given turn.

a 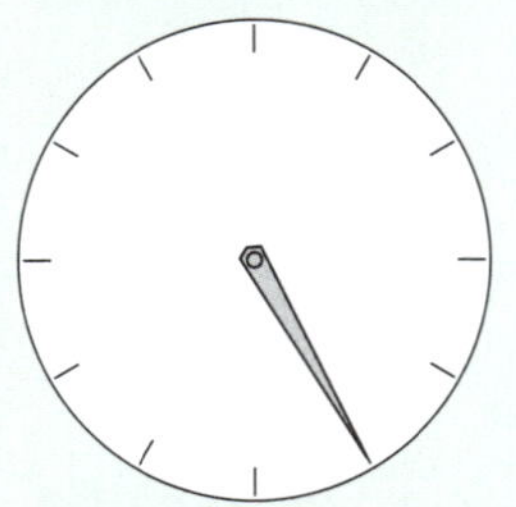

b 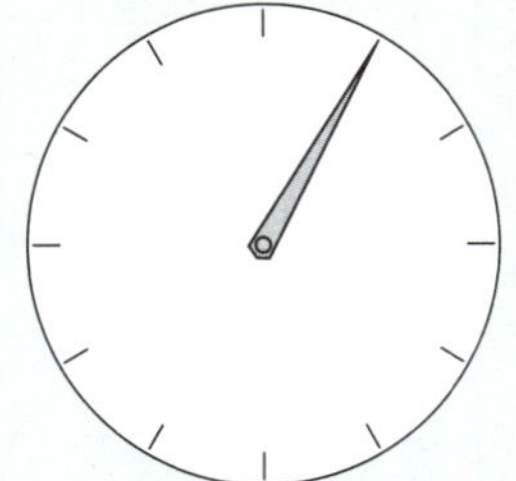

c

quarter turn clockwise   half turn clockwise   quarter turn anticlockwise

Exercise 1

To make sure you are ready to work with money, practise your counting.

a Add 5 each time.

5 → ☐ → ☐ → ☐ → ☐ → 30 → ☐ → ☐

b Add 10 each time.

10 → ☐ → ☐ → ☐ → 50 → ☐ → ☐ → ☐

c Add 20 each time.

20 → ☐ → ☐ → ☐ → ☐

d Add 20 each time.

50 → ☐ → ☐ → ☐ → ☐

e Add 50 each time.

50 → ☐ → ☐ → ☐

We use **money** to buy items.

Different types of money are used in different countries.

Swiss franc **United States dollar**

The money used in a country is called its **currency**.

In most countries, the currency includes **coins** and **notes**.

Exercise 2

Practise your spelling.

money ___________________________   currency ___________________________

___________________________   ___________________________

coins ___________________________   notes ___________________________

___________________________   ___________________________

Coins are mainly used for small amounts of money.

Notes are mainly used for large amounts of money.

Activity

As a class, list as many different currencies from nearby countries as you can.

Swiss franc   United States dollar

___________________________   ___________________________

___________________________   ___________________________

___________________________   ___________________________

___________________________   ___________________________

Click on the icon to print some Exercises for your currency.

CHAPTER 16: CHANCE

We often use words to describe the **chance** of an event happening.

Exercise 1

Circle the word or phrase which describes the *chance* of the event.

a "I am certain the sun will rise tomorrow."

b "It is unlikely to rain this afternoon."

c "The netball team will probably win their match on Friday."

d "I am almost certain my father will pick me up after school today."

e "Our new gym is likely to be ready for the new school year."

If an event will definitely happen, we say it is **certain** to happen.

If an event will definitely not happen, we say it is **impossible** to happen.

An event has **50-50 chance** if it is as likely to happen as not happen.

When I toss a coin, the two possible **outcomes** are "head" and "tail".

I am as likely to toss a "head" as to *not* toss a "head".

There is a 50-50 chance of tossing a head.

head   tail

Exercise 2

Complete:

When I toss a coin, there is a _________________ chance of tossing a tail.

Activity

What to do:

1 Working with a partner, order these words from impossible to certain.

likely

almost certain

very unlikely

50-50 chance

almost impossible

very likely

unlikely

certain

impossible

2 Check your answers with the rest of your class.

Activity

Think ahead to your next weekend. Write down an activity which *for you* is:

almost certain

likely

50-50 chance

unlikely

almost impossible

Discussion

What can we really be *certain* about?

Exercise 3

Use a word or phrase to describe the chance that the spinner will stop on pink.

Remember you can also use *certain* and *impossible*.

a 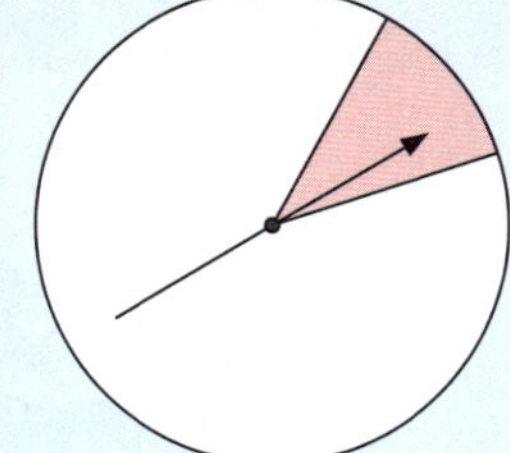

b 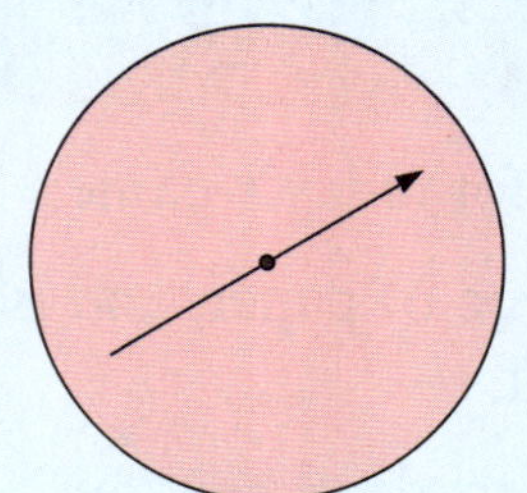

c 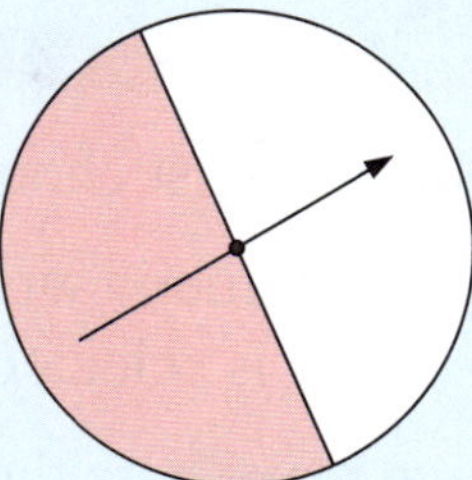

d 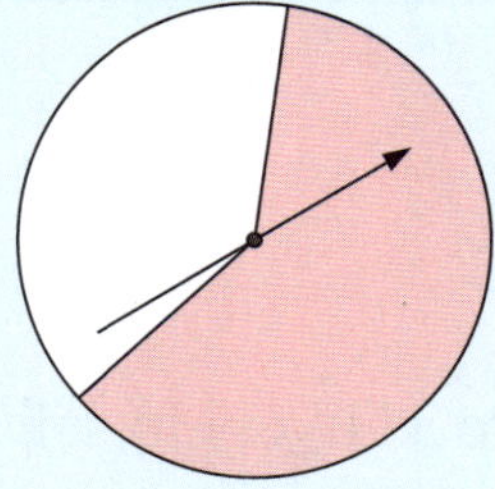

e 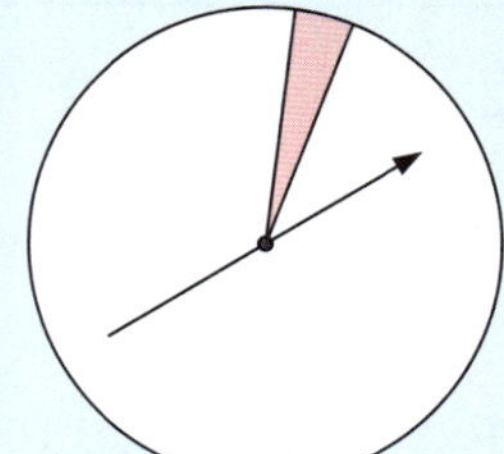

f 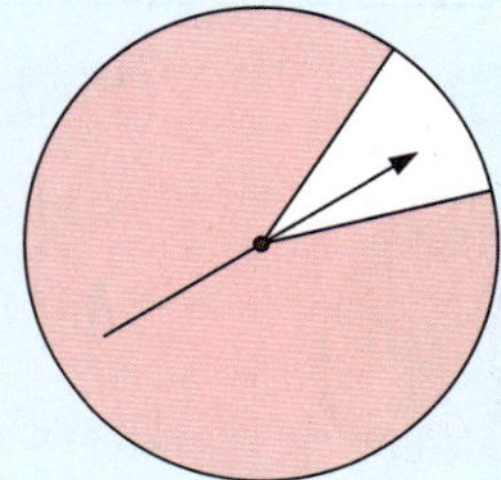

g 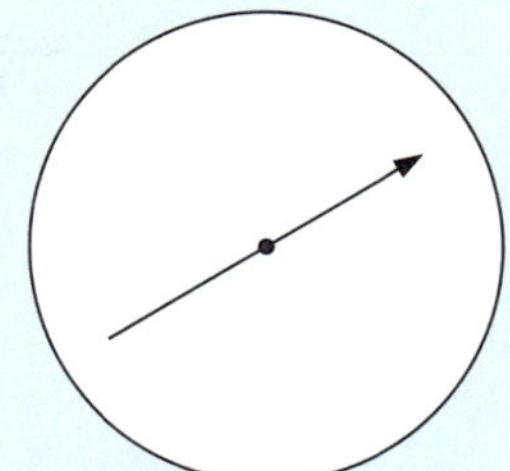

h 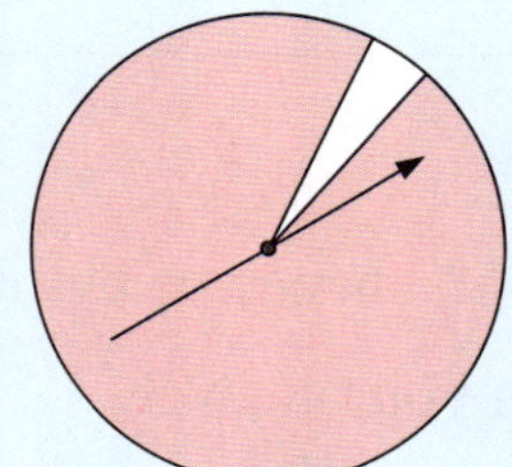

i 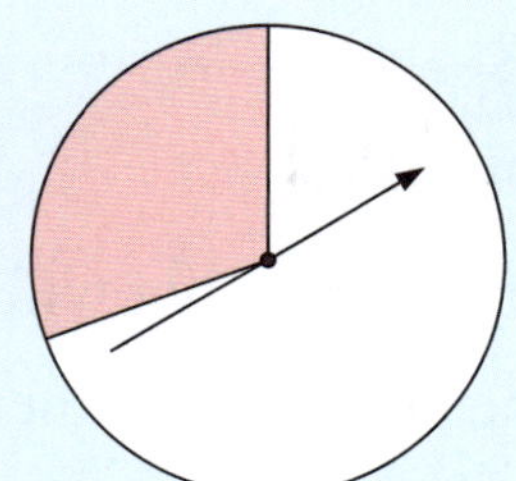

Puzzle

Divide the spinner into four regions and colour them so that:

- there is a 50-50 chance of spinning purple
- there is the same chance of yellow as green
- the least likely colour is brown.

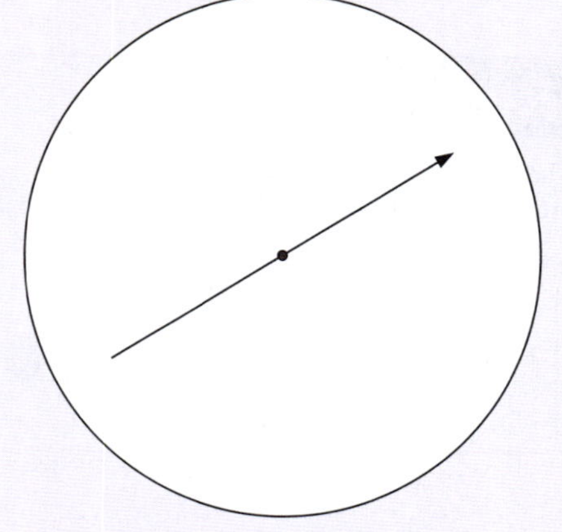

We often talk about "selecting" or "pulling out" an object from a bag or hat *at random*.

In a **random selection**, every object is **equally likely** to be chosen.

Discussion

A football team must choose a captain for today's game.

Each player's name is written on a piece of paper, and placed in a hat. Then, *one* name is selected from the hat.

How can we make sure the selection is made *at random*?

This bag contains 2 red balls and 2 yellow balls.

2 out of 4 balls are red.

2 out of 4 balls are yellow.

If we *randomly select* a ball from the bag, we are as likely to select a red ball as to *not* select a red ball.

There is a 50-50 chance of selecting a red ball.

Exercise 4

From which container will I have a 50-50 chance of pulling out an orange ball?

Circle the correct container and explain your answer.

A    B    C    D 

____

____

Exercise 5

Circle the container that will give a 50-50 chance of pulling out a purple ball.

A    B    C    D 

Exercise 6

Colour enough balls in each bag so the phrase underneath describes the chance of randomly selecting a coloured ball.

a  **b**  **c**  **d** 

    unlikely   certain   very likely   50-50 chance

Exercise 7

This bag contains 100 balls. There are:

- 49 red balls
- 30 blue balls
- 20 green balls
- 1 gold ball.

One ball is randomly selected from the bag.

Match each event with the correct description.

selecting a green ball	50-50 chance
selecting the gold ball	almost certain
selecting a blue ball or a green ball	unlikely
not selecting the gold ball	almost impossible

Exercise 8

a When this spinner is spun, the possible outcomes are

_____________________ , _____________________ , and

_____________________ .

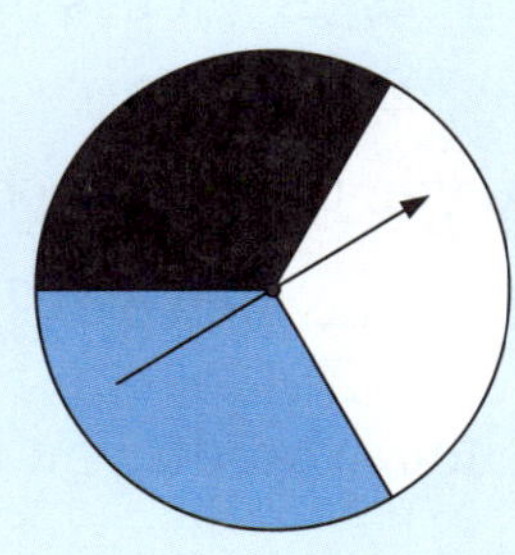

b Are the possible outcomes equally likely? ___________

c Is there a 50-50 chance of getting each outcome? ___________

If not, describe the chance of each outcome. _______________________________

Exercise 9

When I roll a die, the possible outcomes are

________, ________, ________, ________, ________,   and   ________ .

Are the possible outcomes equally likely?   ________

Activity   Rolling a die

You will need: a die

What to do:

1 Roll the die 40 times.

Record each result in the tally.

2 Complete this statement:

In total, I rolled ______ ones, ______ twos,

______ threes, ______ fours, ______ fives,

and ______ sixes.

Outcome	Tally
1	
2	
3	
4	
5	
6	

3 When a die is rolled many times, Jenny thinks each number will be rolled *about* the same number of times.

Jenny is correct because the possible outcomes are

________________________ ________________________ .

Discussion

In their last 20 games of chess, Jada won 17 times, and Simon won 3 times.

1 Do you think each person is *equally likely* to win the next time they play?   ________

2 If not, who do you think is *more likely* to win the next game?   ________

3 Describe the chance that Simon will win the next game.

____

Activity

You will need: two identical coins

When two identical coins are tossed, we could get:

- two heads
- two tails
- a head and a tail.

What to do:

1 Toss 2 coins 40 times. Record the result for each outcome using a tally.

Outcome	Tally

2 Compare your results with your classmates.

   a Did everyone get *exactly* the same results? _______________

   b Describe the general *pattern* of most people's results.

   ____

   ____

3 Do you think the different outcomes are equally likely? Explain your answer.

____

____

Discussion

- At the start of a cricket match, the captains of two teams toss a coin to see which team will bat first.

  Why is tossing a coin considered *fair*?

- Belinda and her two friends Sally and Theresa have only two tickets for a roller coaster.

  How could the girls use an ordinary die to fairly decide who misses out on the ride?

- Five friends are eating a box of chocolates.

  When everyone has had 3 chocolates each, there is 1 chocolate remaining. How could the friends use an ordinary die to fairly decide who gets the last chocolate?

Activity   Coin toss race

Click the icon to open the game. The game should be played in pairs.

The game will tell you when to answer the questions below.

1 Who won the race around the world? ___________________________

2 How many times did the coin land on: **a** heads _________   **b** tails? _________

3 If you played again, do you know whether the same person would win? _________

I think this is because _______________________________________

____

Revision

1 Choose the word or phrase which best describes the chance that each spinner will stop on blue.

| unlikely | certain | 50-50 chance | very likely | almost impossible |

a

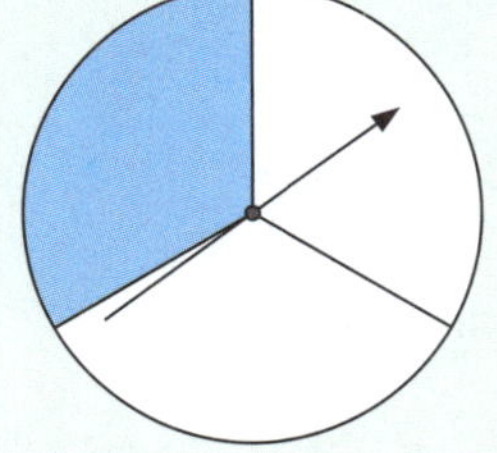

b

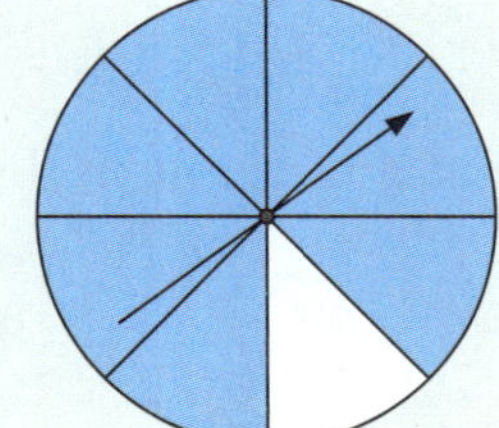

c

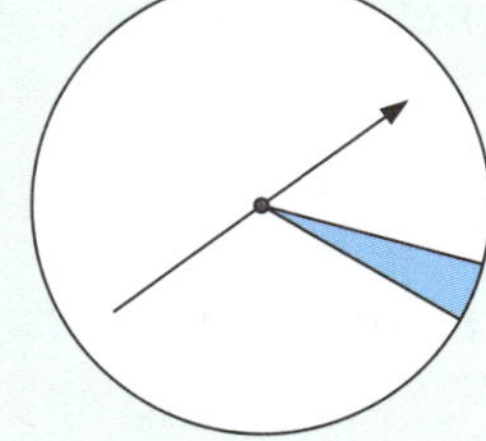

_______________   _______________   _______________

d

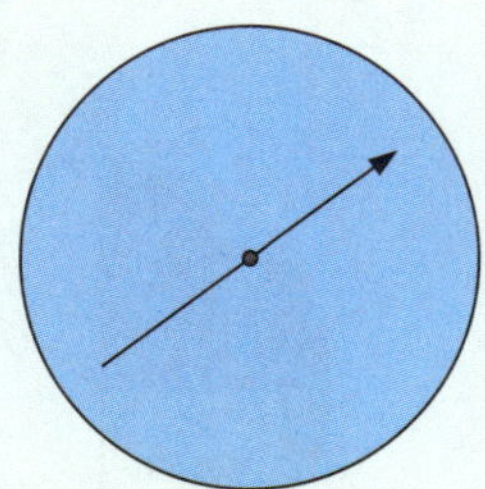

e

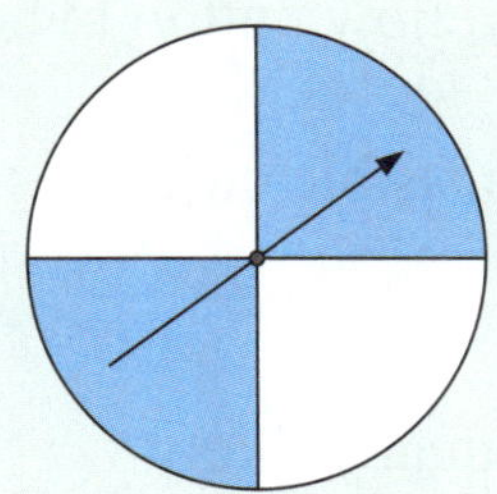

_______________   _______________

2 Complete:

In a *random selection*, every object is _______________ _______________
to be chosen.

3 Circle the container that will give a 50-50 chance of randomly selecting a green ticket.

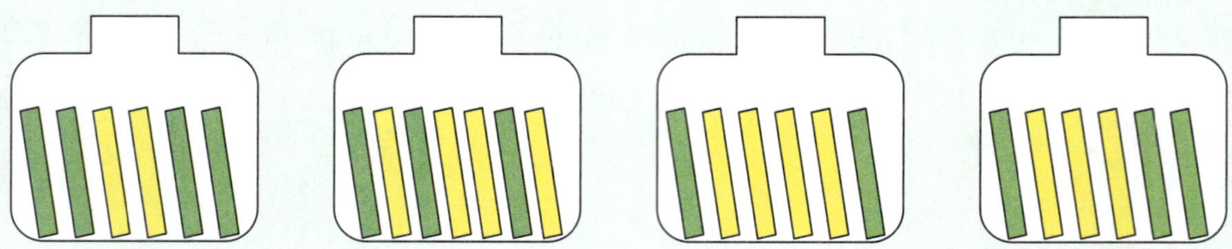

4 Colour enough balls in each bag so the phrase underneath describes the chance of randomly selecting a coloured ball.

a 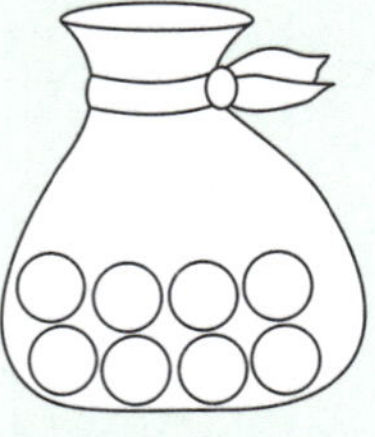

50-50 chance

b 

very unlikely

c 

likely

5 This vase contains 50 buttons. There are:

- 40 red buttons
- 9 green buttons
- 1 black button.

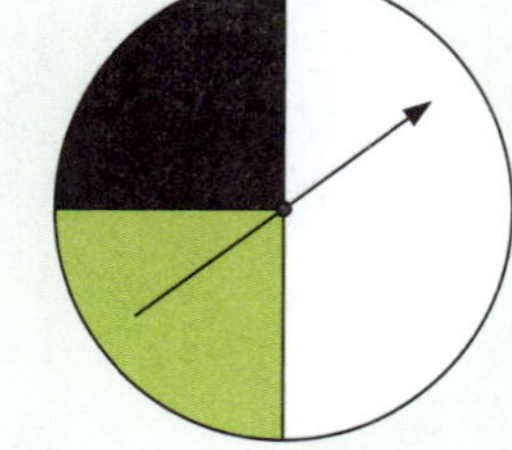

One button is randomly selected from the vase.

Complete:

a It is very _________________ that a red button will be selected.

b It is almost impossible for the _______________ button to be selected.

c It is _________________ _________________ that a red button or a green button will be selected.

d It is _______________ for a yellow button to be selected.

6 a When this spinner is spun, the possible outcomes are

_______________, _______________, and _______________.

b Are the possible outcomes *all* equally likely? _________

c Which outcome has a 50-50 chance of happening?

____

d Which *two* outcomes are equally likely to happen?

_______________ and _______________.

CHAPTER 17: TRANSFORMATIONS

Jamal can copy the shape and **move** it to the right of the previous shape.

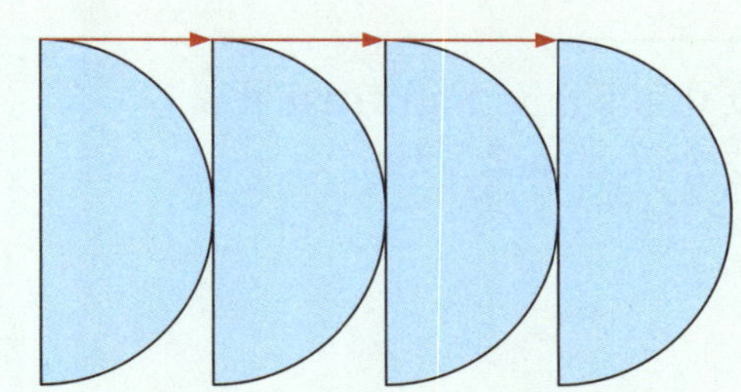

Exercise 1

Draw *three* more identical shapes to complete each pattern:

a

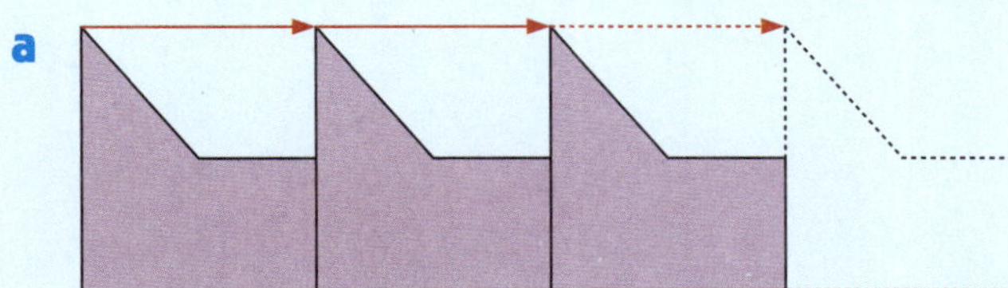

b

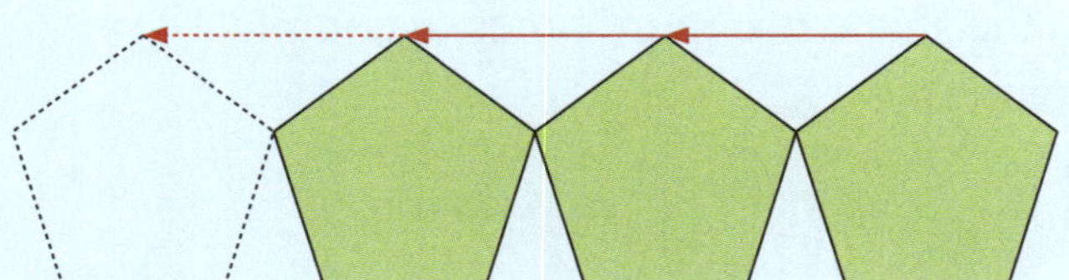

c

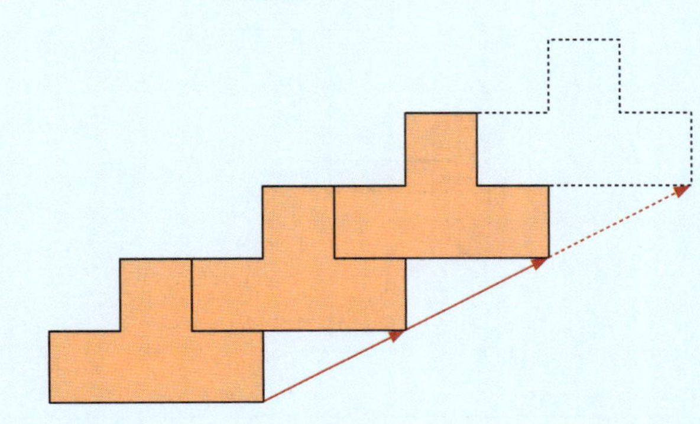

Activity

Make a pattern of your own by copying and moving this shape.

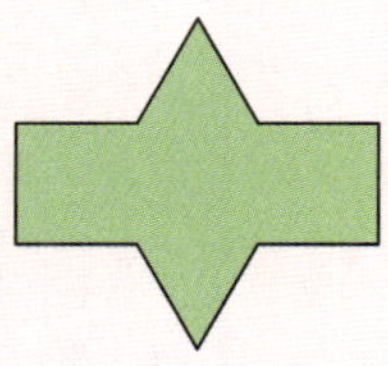

Jamal can copy this shape and **turn** it a *quarter turn clockwise* each time.

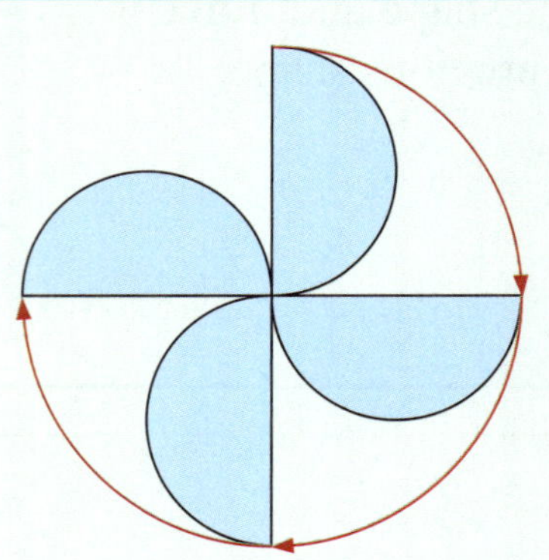

Exercise 2

a This is a quarter turn in a

_________________________ direction.

Turn the shape *two* more times to complete the pattern.

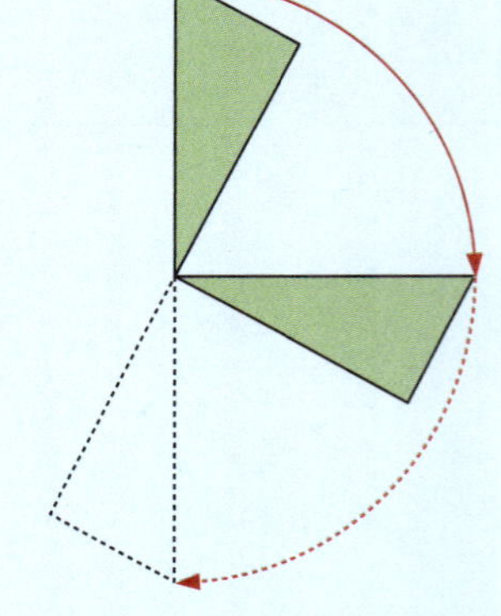

b This is a quarter turn in an

_________________________ direction.

Turn the shape *two* more times to complete the pattern.

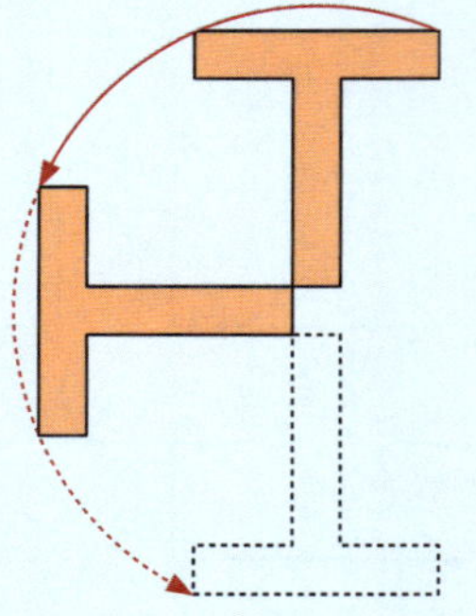

Activity

- Which of these shapes is formed by *moving* the blue shape? Colour these shapes in red.
- Which of these shapes is formed by *turning* the blue shape a quarter turn *clockwise*? Colour these shapes in green.
- Which of these shapes is formed by *turning* the blue shape a quarter turn *anticlockwise*? Colour these shapes in yellow.

Jamal cut his shape out and held it against a mirror.

The resulting shape is a **reflection** of the original shape.

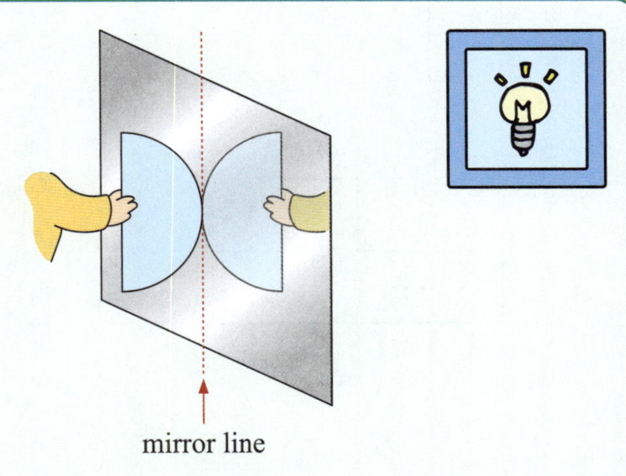

Discussion

Where do you see reflections?

What kinds of objects behave as mirrors?

____

____

____

____

Discussion

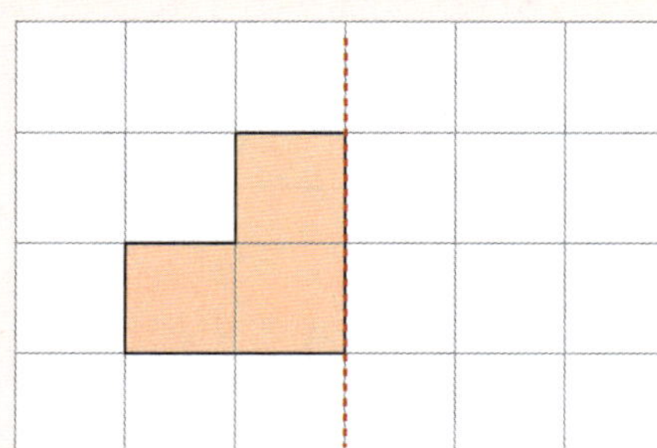

How can you *reflect* this figure in the red *mirror line*?

Draw the reflection together.

Exercise 3

Reflect each figure in the mirror line.

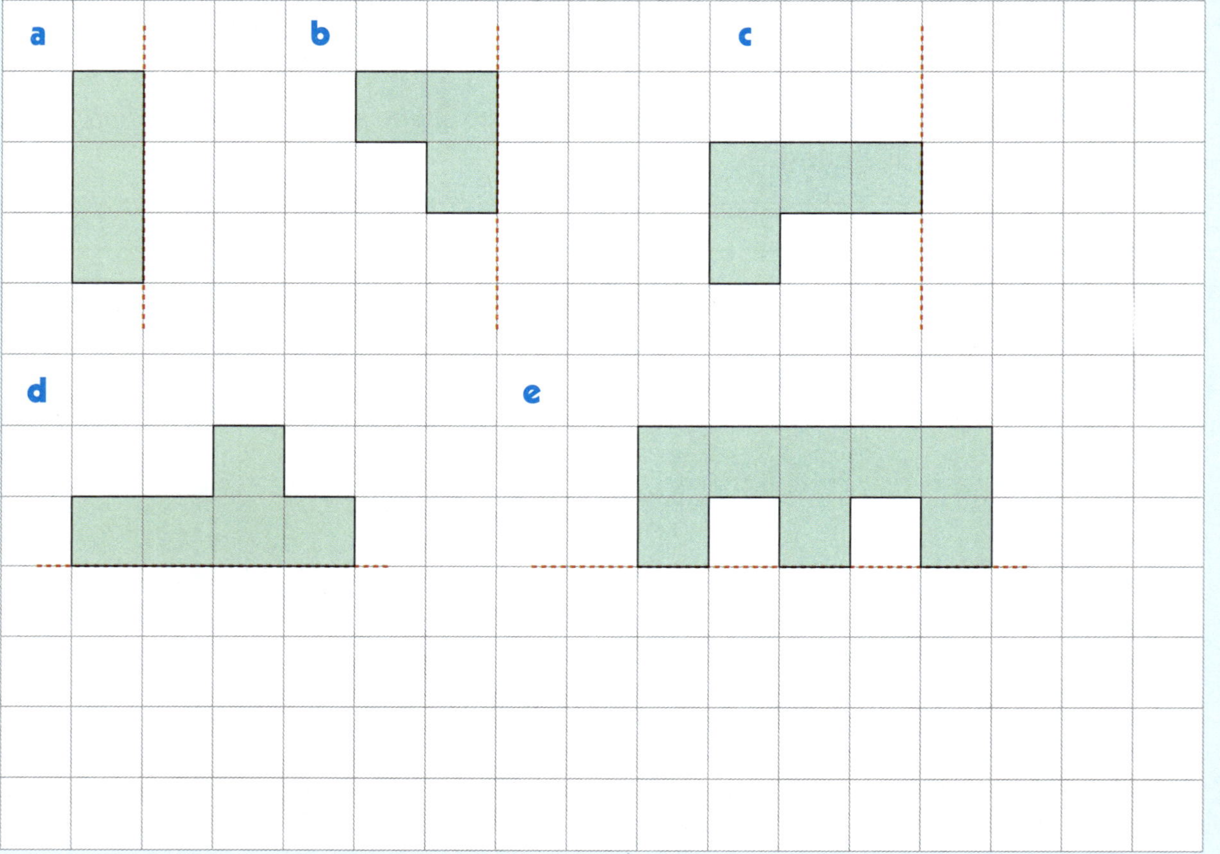

Activity

Stand in front of a mirror, and lift your right hand.

Does it *look* like your right hand is lifted?

Discussion

When we *move*, or *turn*, or *reflect* an object, does the object change *shape*? _______________

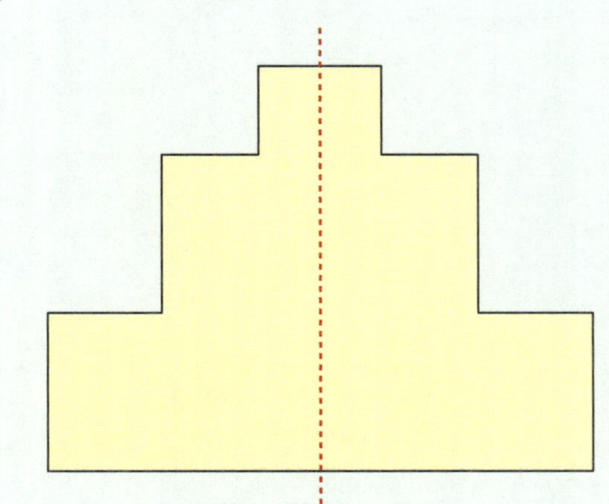 

The parts of this figure either side of the red line are mirror images of each other.

We say that the figure is **symmetrical**.

The red line is called **line of symmetry**.

We can use a mirror to show that one side is a **reflection** of the other side.

Exercise 4

For each figure, place a ✓ next to the correct line of symmetry.

a 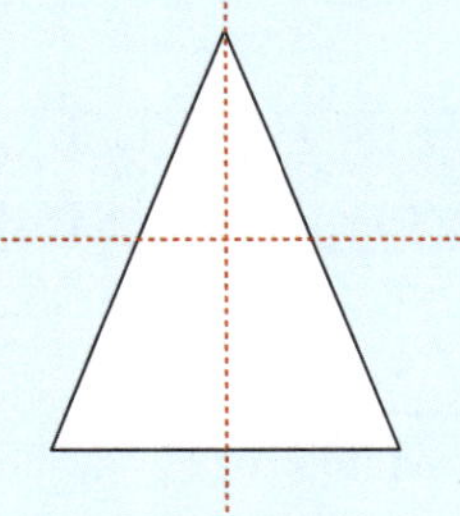   **b** 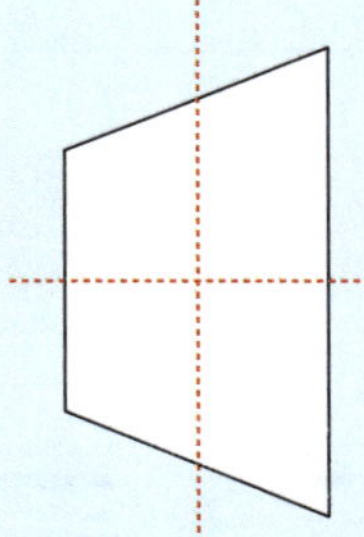   **c** 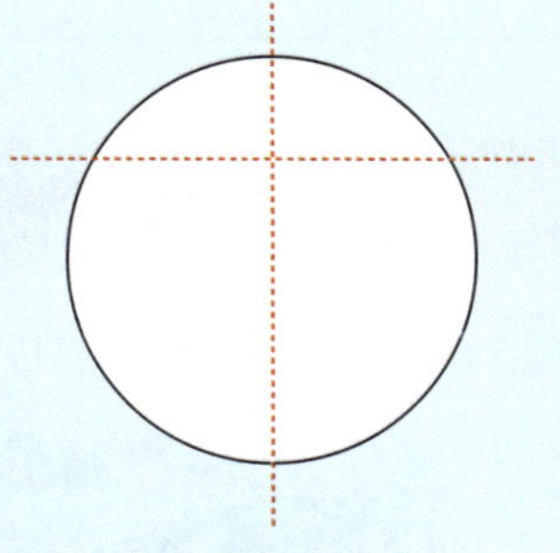

Puzzle

Circle the house that *is* symmetrical. Draw its line of symmetry.

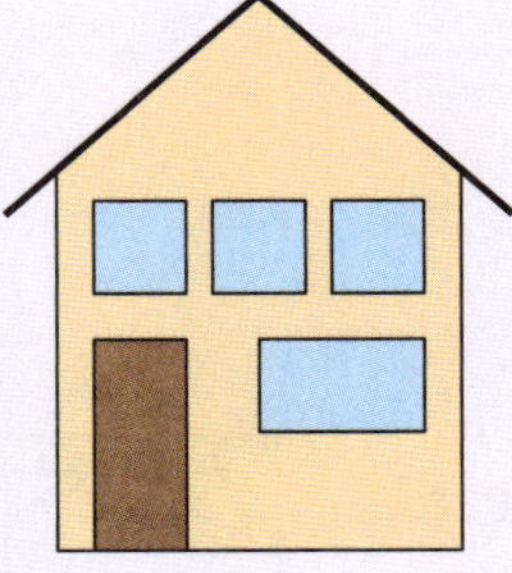 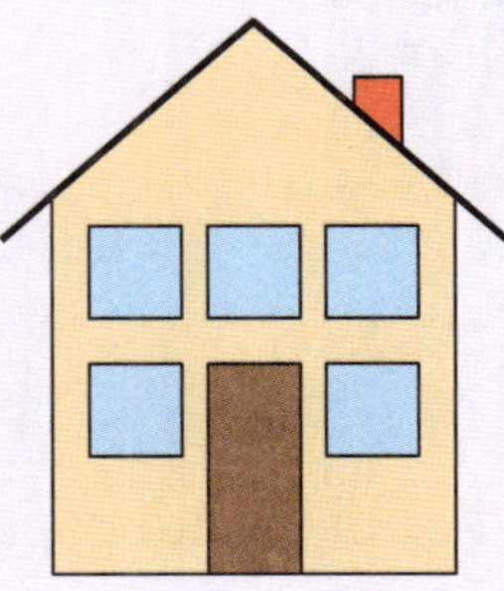 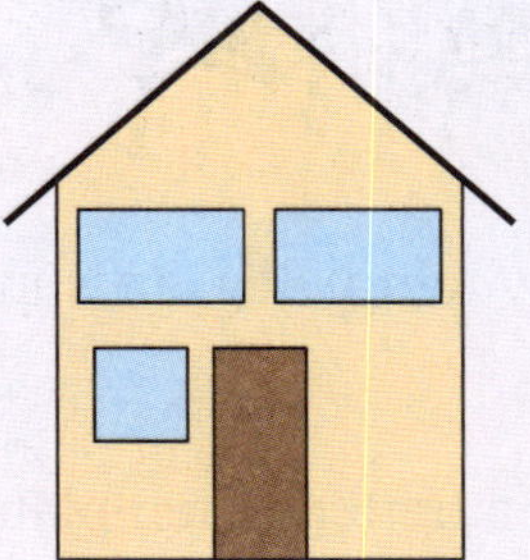 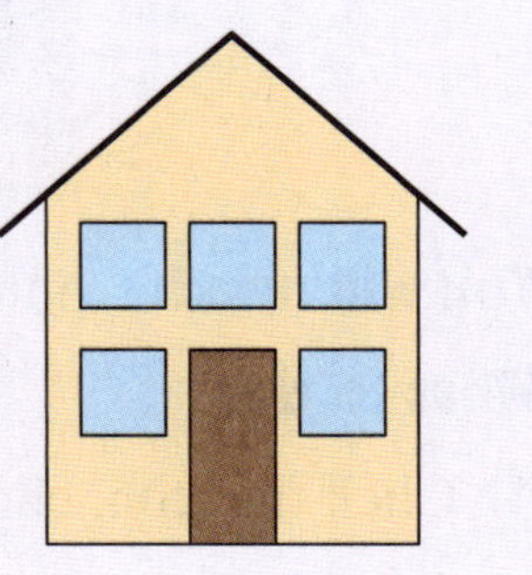

Exercise 5

Use a ruler to draw any lines of symmetry for these figures.

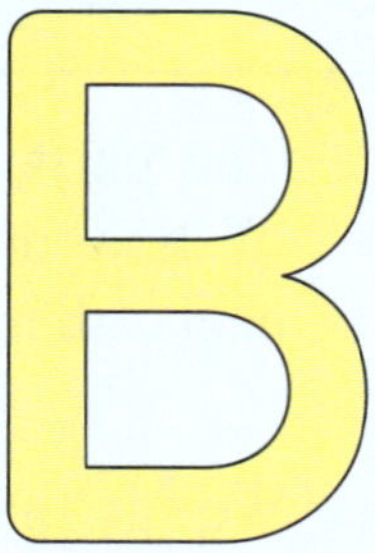

a T

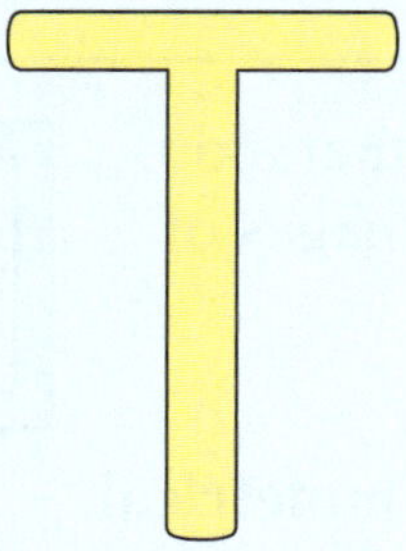

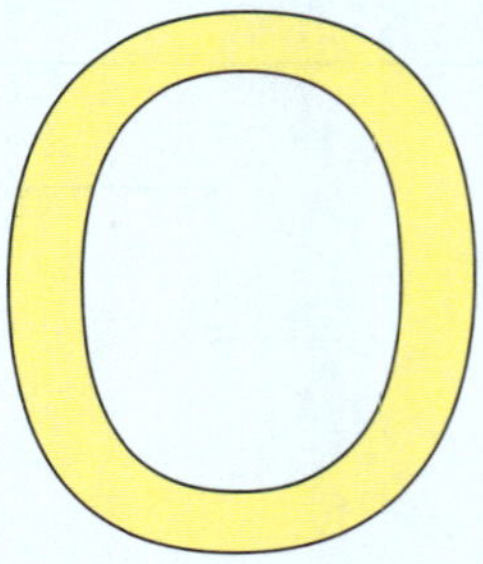

b B

c O

d W

e X

Activity   Butterflies

The wings of a butterfly are usually symmetrical.

You will need: butterfly templates, art supplies

What to do:

1 Click the icon, choose a butterfly template, and print it onto paper or card.

2 Use a pencil to create a symmetrical pattern on the wings.

3 Colour your butterfly.

4 Carefully cut around your butterfly, and hand it to your class teacher for display.

Activity — Snowflakes

If you place a single snowflake under a microscope, you might see a beautiful symmetrical figure.

You will need: white paper squares, coloured paper, scissors, glue, tape

What to do:

1 Click the video clip. Watch how to make a snowflake by cutting and folding your white paper squares.

You should create *four* snowflakes.

2 Arrange your snowflakes onto the coloured paper. Glue them down when you are happy with your arrangement.

Activity

Many leaves and flowers show symmetry. They can be used to create symmetrical designs.

Collect items from your environment to create a piece of symmetrical art.

We can sometimes use copies of a shape to make a pattern which completely covers a surface, with no gaps.

- We can use these identical squares to cover the surface with no gaps.

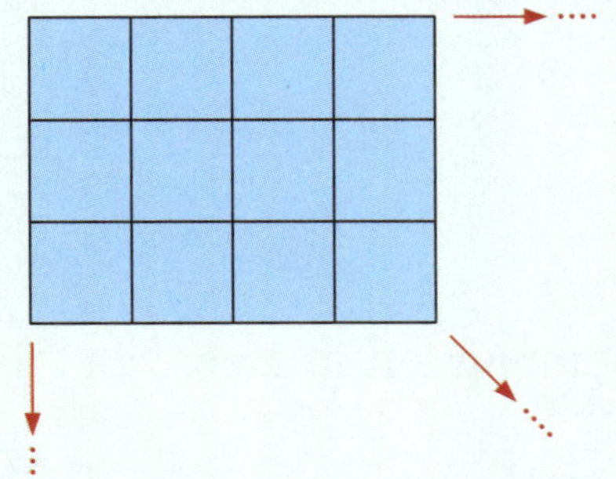

- We *cannot* use these identical pentagons to cover the surface. There are gaps between the shapes.

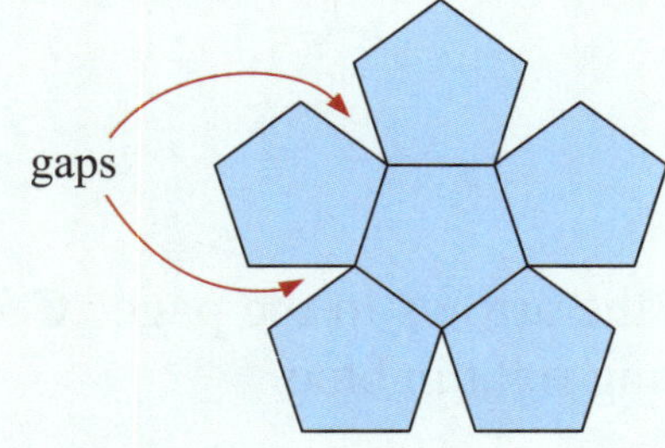

Exercise 6

a Draw more identical hexagons to cover the white surface, with no gaps.

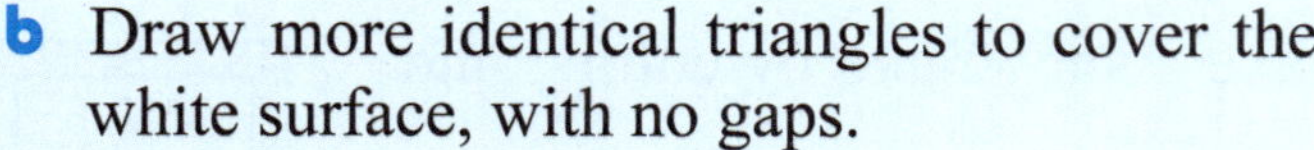

b Draw more identical triangles to cover the white surface, with no gaps.

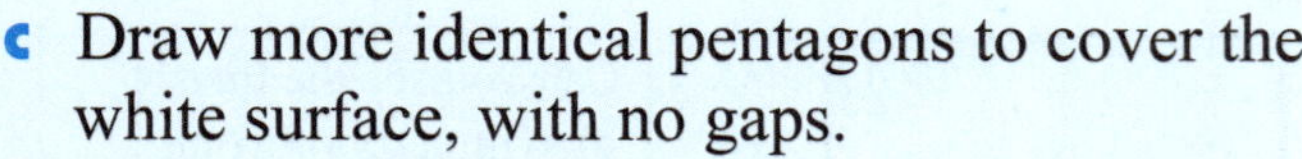

c Draw more identical pentagons to cover the white surface, with no gaps.

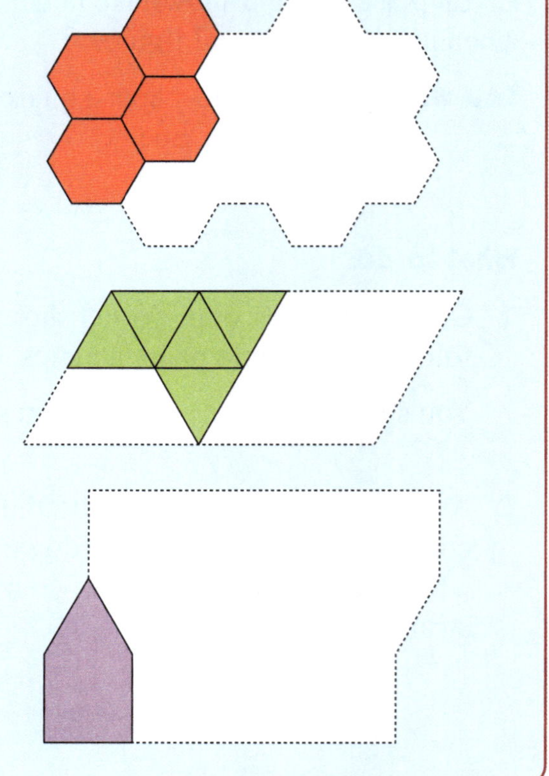

Activity   Arrows

You will need: scissors, glue, coloured pencils

What to do:

1 Click the icon.

Print the 12 arrows, then cut them out.

2 Arrange the arrows on the page so they completely cover the white surface, with no gaps. Six of your arrows should point *left*, and six arrows should point *right*.

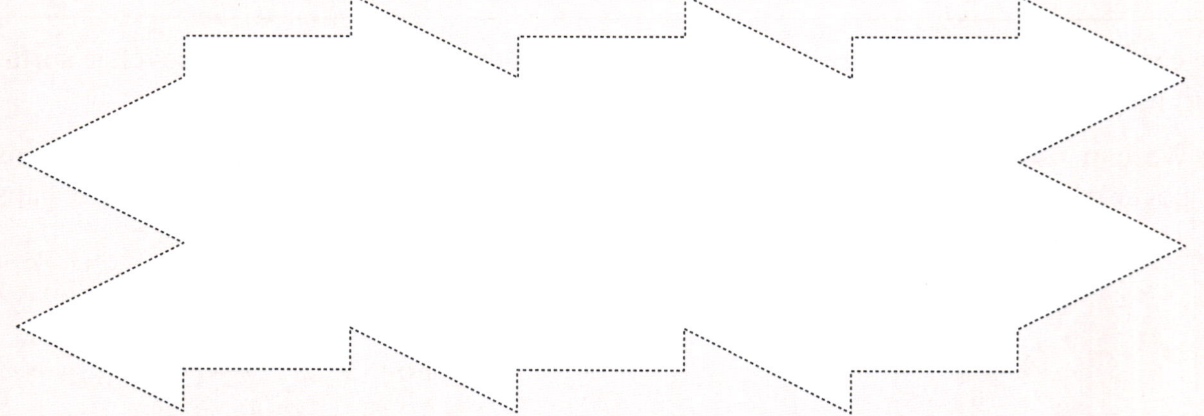

3 Glue the arrows to the page. Colour the arrows pointing left in green, and the arrows pointing right in blue.

Activity

Patterns of shapes that cover surfaces can be seen in the world around us.

What to do:

1 Write down the shape used in:

    a a bee's honeycomb

    b these bricks.

2 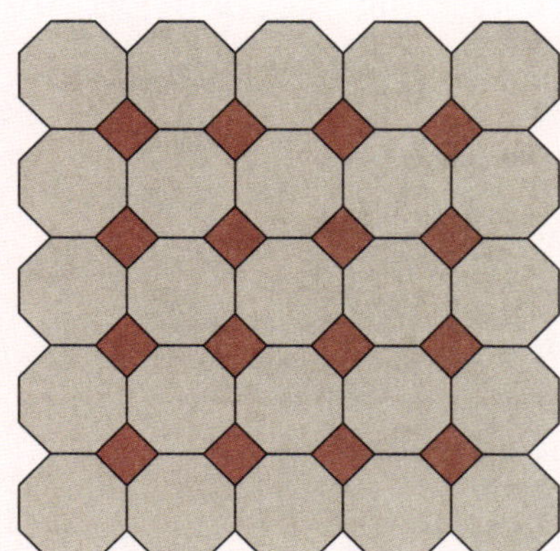

Write down the *two* shapes used in this pattern of floor tiles.

3 Draw a floor tile pattern of your own, which uses at least two different shapes to cover the surface.

Revision

1   Draw *three* more identical shapes to complete this pattern.

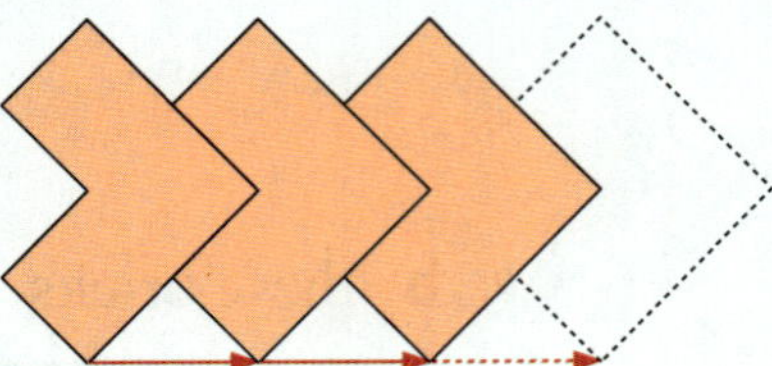

2   This is a quarter turn in a

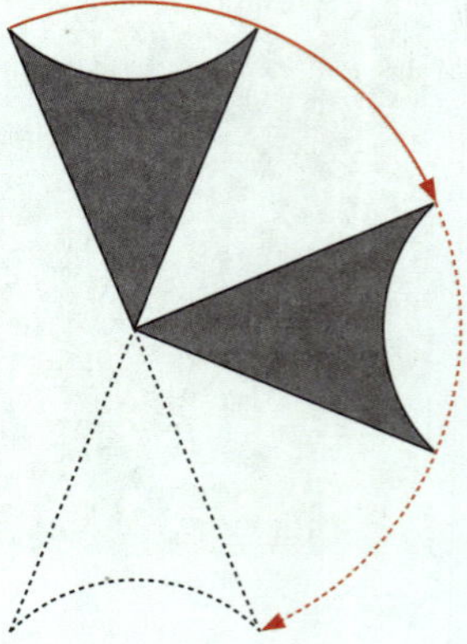

__________________________ direction.

Turn the shape *two* more times to complete the pattern.

3   Reflect each figure in the mirror line.

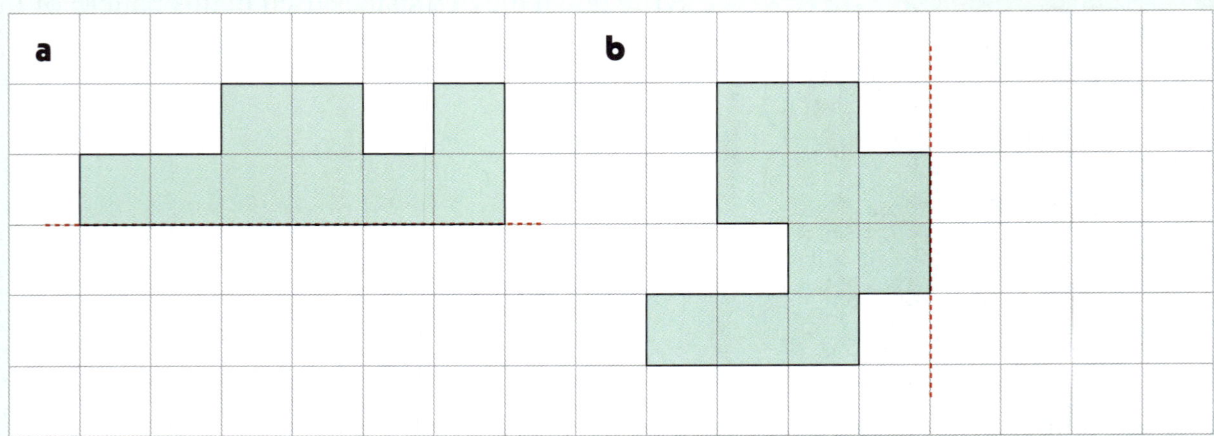

4   Use a ruler to draw any lines of symmetry for these figures.

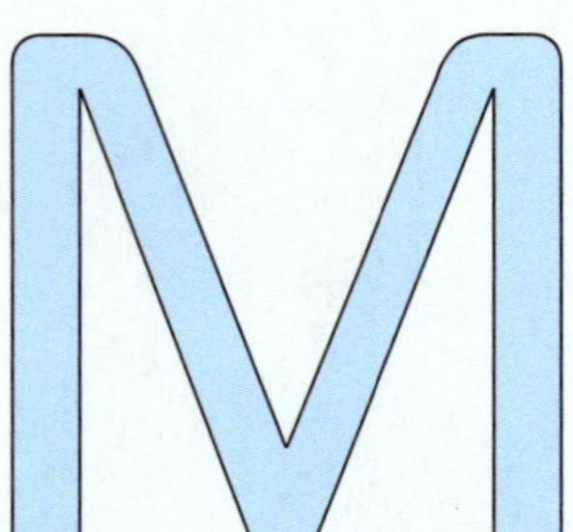
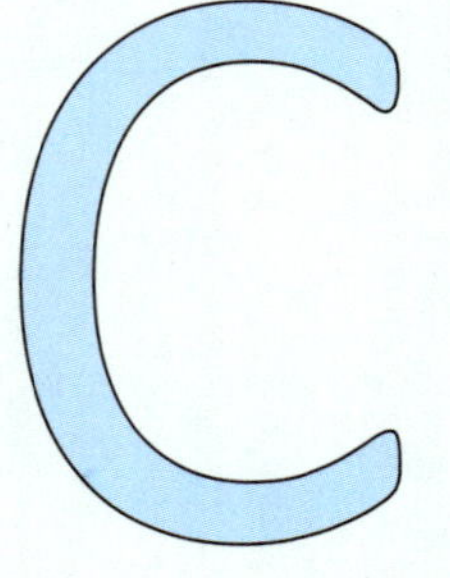
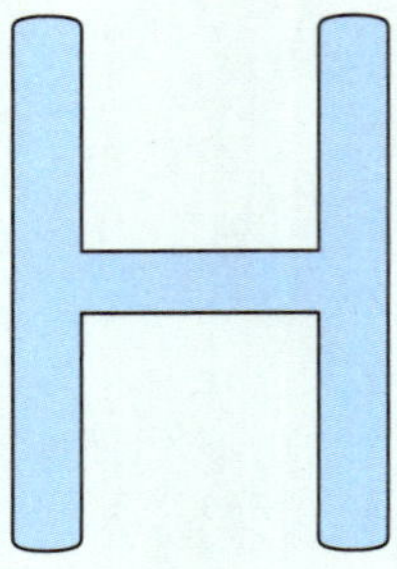

5   Draw more identical quadrilaterals to cover the white surface, with no gaps.

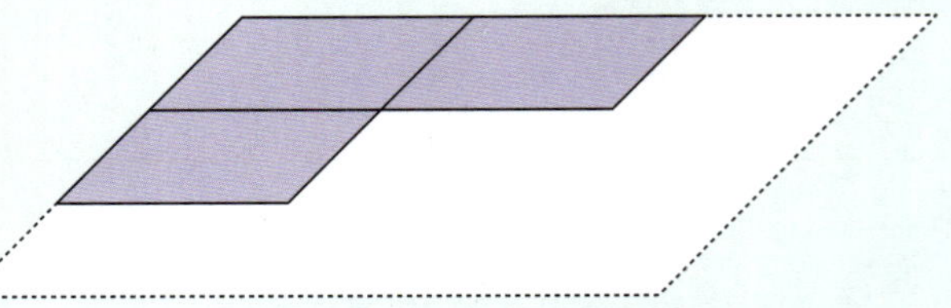